28

CONTEMPORARY'S
REAL NUMBERS
Developing Thinking Skills in Math
Measurement

Allan D. Suter

Project Editor
Kathy Osmus

CB
CONTEMPORARY BOOKS

a division of NTC/Contemporary Publishing Group
Lincolnwood, Illinois USA

ISBN: 0-8092-4208-7

Published by Contemporary Books,
a division of NTC/Contemporary Publishing Group, Inc.,
4255 West Touhy Avenue,
Lincolnwood (Chicago), Illinois, 60712-1975 U.S.A.

1 2 3 4 5 6 C(K)   20 19 18 17 16 15 14

Editorial Director	Production Editor
Caren Van Slyke	Marina Micari

Editorial	Cover Design
Lisa Black	Lois Stein
Janice Bryant	
Lisa Dillman	Illustrator
Ellen Frechette	Cliff Hayes
Leah Mayes	
Steve Miller	Art & Production
Robin O'Connor	Sue Springston
Betsy Rubin	Rosemary Morrissey-Herzberg
Seija Suter	
	Typography
Editorial Production Manager	J•B Typesetting
Norma Fioretti	St. Charles, Illinois

Cover photo © by Michael Slaughter

CONTENTS

Scales in Everyday Life

A scale is a series of evenly spaced numbers used to measure.

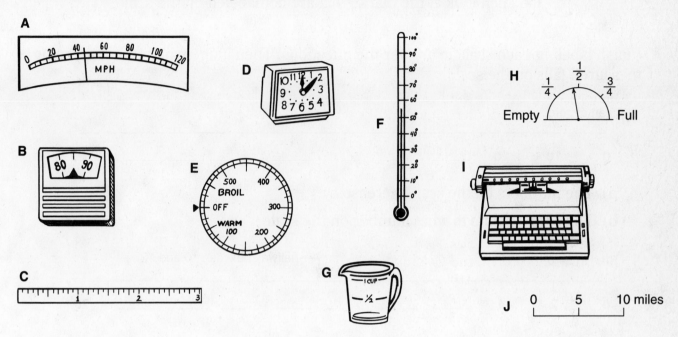

▶ Decide which scale you would use to measure each item. Then write the letter of the scale you chose in the space provided.

1. The speed of a car __A__
letter

2. Your height _____
letter

3. Set the margin for typing a letter _____
letter

4. The time of day _____
letter

5. Your weight in pounds _____
letter

6. Amount of liquid needed _____
letter

7. Outside temperature _____
letter

8. The amount of gasoline in a car _____
letter

9. Oven temperature _____
letter

10. Estimate the distance on a map _____
letter

Reading Scales

To read a scale: • Look at the numbers that label the scale.

  • Then look at the marks that are not labeled on the scale.

▶ Finish labeling each scale to answer the questions. Then judge where the item being measured falls on the scale.

1. Start at zero.

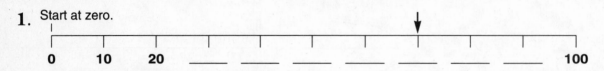

a) The numbers on the scale increase by __10__ .

b) The arrow points to what number on the scale? _____

2. Start at zero.

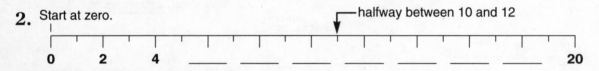

a) The numbers on the scale increase by __2__ .

b) The arrow points to what number on the scale? _____

3. Start at zero.

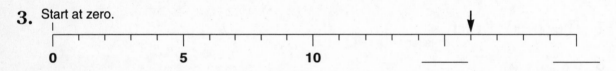

a) The numbers on the scale increase by _____ .

b) The arrow points to what number on the scale? _____

4. Start at zero.

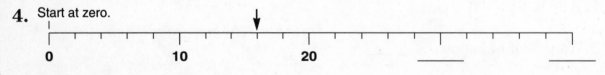

a) The numbers on the scale increase by _____ .

b) The arrow points to what number on the scale? _____

5. Start at zero.

a) The numbers on the scale increase by _____ .

b) The arrow points to what number on the scale? _____

Estimate Using Scales

Sometimes you will need to read a scale between two markings. Then you will need to judge the approximate value.

Example A

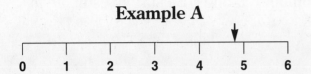

The reading is between 4 and 5 but closer to 5. The reading is about 5.

Example B

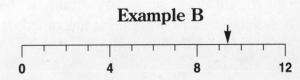

The reading is between 9 and 10 but closer to 9. The reading is about 9.

▶ For each scale, estimate the reading to the nearest whole number.

1.

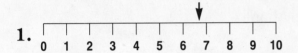

    a) The reading is between __6__ and _____ .

    b) The reading is closer to _____ .

    c) The reading is about _____ .

2.

    a) The reading is between _____ and _____ .

    b) The reading is closer to _____ .

    c) The reading is about _____ .

3.

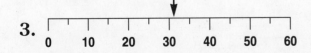

    a) The reading is between _____ and _____ .

    b) The reading is closer to _____ .

    c) The reading is about _____ .

4.

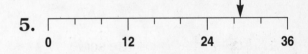

    a) The reading is between _____ and _____ .

    b) The reading is closer to _____ .

    c) The reading is about _____ .

5.

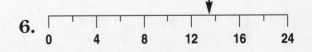

    a) The reading is between _____ and _____ .

    b) The reading is closer to _____ .

    c) The reading is about _____ .

6.

    a) The reading is between _____ and _____ .

    b) The reading is closer to _____ .

    c) The reading is about _____ .

Reading Different Scales

The ability to read different kinds of scales accurately is an important life skill.

▶ Use the different scales to estimate whole number readings. Be sure to estimate to the nearest mark whether it is labeled or not.

1.

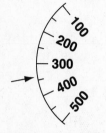

About ___350___

2.

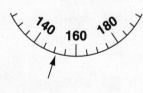

About _____

3.

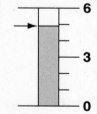

About _____

4.

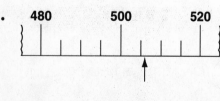

About _____

5.

About _____

6.

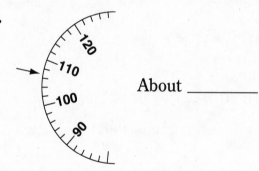

About _____

7.

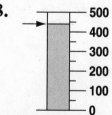

About _____

8.

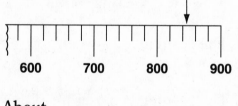

About _____

Scales with Decimals

Sometimes scales measure in tenths. The line below has been divided into ten equal parts. Each part is one tenth of the line.

Example

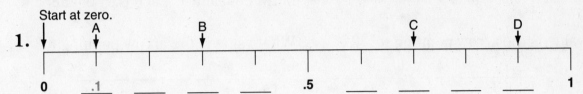

0   .1   .2   .3   .4   .5   .6   .7   .8   .9   1

▶ Finish labeling each scale. Then find the value for each letter, and write it in the space provided.

1. Start at zero.

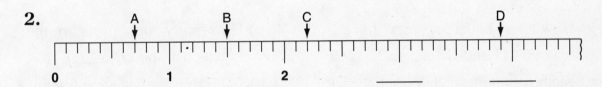

0   .1 ___ ___ .5 ___ ___ ___ ___ 1

a) Value for letter A ___.1___   **c)** Value for letter C _____

b) Value for letter B _____   **d)** Value for letter D _____

2.

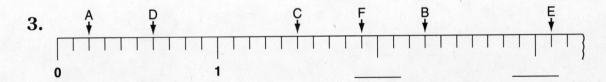

0   1   2   _____   _____

a) Value for letter A _____   **c)** Value for letter C _____

b) Value for letter B ___1.5___   **d)** Value for letter D _____

3.

0   1   _____   _____

a) Value for letter A _____   **d)** Value for letter D _____

b) Value for letter B _____   **e)** Value for letter E _____

c) Value for letter C _____   **f)** Value for letter F _____

Reading Scales with Fractions

To read a fraction on a scale, you need to know the size of the smallest parts that the scale is divided into.

Example A

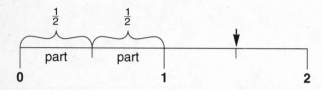

The scale shows two equal parts between 0 and 1. Each part equals $\frac{1}{2}$.

What value is the arrow pointing to? $1\frac{1}{2}$

Example B

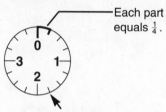

Each part equals $\frac{1}{4}$.

The scale shows four equal parts between 0 and 1. Each part equals $\frac{1}{4}$.

What value is the arrow pointing to? $1\frac{3}{4}$

▶ Answer the questions to help you read each scale.

1. a) How many equal parts are there between 11 and 12? ___4___

b) Each part equals what fraction? _____

c) What value is the arrow pointing to? _____

3. a) How many equal parts are there between 1 and 2? _____

b) Each part equals what fraction? _____

c) What value is the arrow pointing to? _____

2. a) Each part equals what fraction? _____

b) What value is the arrow pointing to? _____

4. a) Each part equals what fraction? _____

b) What value is the arrow pointing to? _____

Everyday Scales

▶ Scales are used to measure many things. Read the scales to estimate your answers.

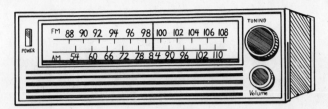

1. **a)** About what setting is shown on the FM scale? _____

   b) About what setting is shown on the AM scale? _____

   c) Place an arrow on the FM scale to mark your favorite station.

2. About how many miles per hour does the speedometer show?

   _____ miles per hour

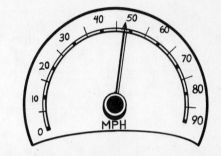

3. Use the map scale to estimate the miles from Shelby to Hart.

   _____ miles

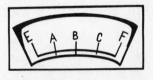

4. What letter on the gas gauge shows the tank is about $\frac{3}{4}$ full?

   _____

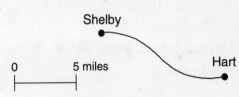

5. About what temperature does the thermometer show?

   _____ degrees

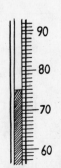

Review Your Skills

▶ Use your skills to answer the following questions.

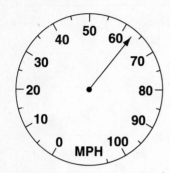

1. About how many miles per hour does the speedometer show?

_____ mph

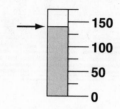

4. Estimate to the nearest mark.

About _____

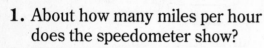

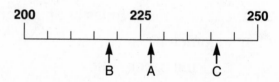

5. Which letter shows:

   a) about 228? _____

   b) about 219? _____

   c) about 240? _____

2. a) The labels on the scale increase by _____ .

   b) Each unlabeled mark shows an increase of _____ .

   c) The arrow points to what number on the scale? _____

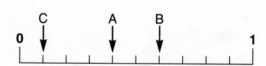

6. Write the decimal and fraction for each letter. Simplify the fraction when necessary.

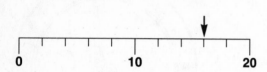

a) Letter A _____ decimal _____ fraction

3. Write the values for each letter.

   a) Value for letter A _____

   b) Value for letter B _____

   c) Value for letter C _____

a) Letter A _____ (decimal) _____ (fraction)

b) Letter B _____ (decimal) _____ (fraction)

c) Letter C _____ (decimal) _____ (fraction)

Measuring Air Pressure

1. The recommended tire pressure for Joan's bicycle tire is 55 pounds.

    a) About what was Joan's first reading? _____ pounds of pressure

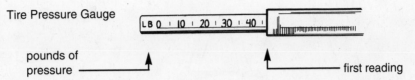

Tire Pressure Gauge

pounds of pressure ⟶ ↑   ↑ ⟶ first reading

    b) After inflating the tire, about what was her second reading? _____ pounds of pressure

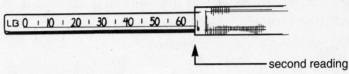

↑ ⟶ second reading

    c) About how many pounds of pressure must be removed to reach the recommended tire pressure? _____

2. The recommended tire pressure for Steve's truck tire is 48 pounds.

    a) About how much was the first reading? _____ pounds of pressure

    b) About how much was the second reading? _____ pounds of pressure

    c) About how many pounds of pressure must be removed to reach the recommended tire pressure? _____

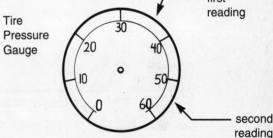

3. The recommended tire pressure for Julie's car is 32 pounds.

    a) About how much was the first reading? _____ pounds of pressure

    b) About how much was the second reading? _____ pounds of pressure

    c) About how many pounds of pressure must be added to reach the recommended tire pressure? _____

Nonstandard Units

Long ago, people used parts of their bodies as general units of measure.

Foot   Span   Pace   Hand   Cubit

The above measures are all nonstandard units since the size of body parts varies from person to person.

▶ Measure the length below using the width of your thumb as your nonstandard unit of measure.

Thumb

Length

1. The length of the arrow is about how many thumb widths? _____

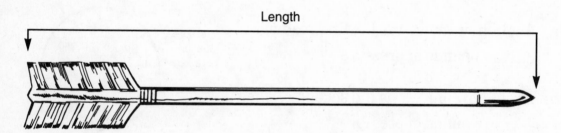

2. The length of the nail is about how many thumb widths? _____

3. The length of your pencil is about how many thumb widths? _____

Standard Units of Length

> To measure is to make a comparison.

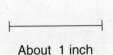

About 1 inch

12 inches is about the length of a man's shoe.

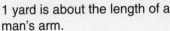

1 yard is about the length of a man's arm.

It takes about 20 minutes to walk 1 mile.

Standard units of length were developed so a given measurement would be the same size to everyone. Without standard units of length, a given measurement does not mean the same thing to all people. A standard unit of measure is always the same size.

Standard Units of Length		
12 inches	=	1 foot
3 feet	=	1 yard
5,280 feet	=	1 mile

▶ Choose the most sensible unit of measure for each problem.

1. The distance from New York to Chicago

   a) inches   **b)** feet   **c)** yards   **d)** miles

2. The length of a paper clip

   a) inches   **b)** feet   **c)** yards   **d)** miles

3. The height of a man or woman

   a) inches   **b)** feet   **c)** yards   **d)** miles

4. The length of a swimming pool

   a) inches   **b)** feet   **c)** yards   **d)** miles

Estimate the Lengths

> The smaller the unit of measure, the closer the estimate.

Example A	Example B
To the nearest inch (unit of measure is 1 inch)	**To the nearest $\frac{1}{2}$ inch** (unit of measure is $\frac{1}{2}$ inch)

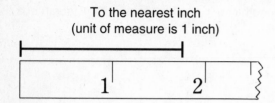 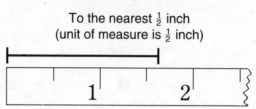

The measure is between 1 and 2 inches but closer to 2 inches.

The measure is between $1\frac{1}{2}$ and 2 inches but closer to $1\frac{1}{2}$ inches.

Estimate: 2 inches

Estimate: $1\frac{1}{2}$ inches

▶ Estimate the lengths.

To the nearest inch:

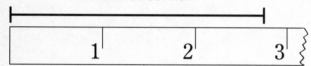

To the nearest $\frac{1}{2}$ inch:

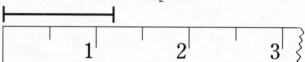

1. a) The measure is between
___2___ and ___3___ inches.

b) The measure is closer to
_____ inches.

c) Estimate: _____ inches

3. a) The measure is between
_____ and ___$1\frac{1}{2}$___ inches.

b) The measure is closer to
_____ inch(es).

c) Estimate: _____ inch(es)

To the nearest inch:

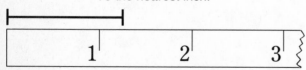

To the nearest $\frac{1}{2}$ inch:

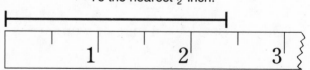

2. Estimate: _____ inch(es)

4. Estimate: _____ inches

Measuring Real-Life Objects

▶ Use the scale as a guide to measure each object to the nearest $\frac{1}{2}$ inch.

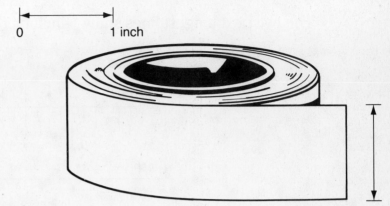

0   1 inch

1. The masking tape is about _____ inch(es) wide.

2. The bolt is about _____ inch(es) long.

3. The nail is about _____ inch(es) long.

4. The button is about _____ inch(es) wide.

5. The screw is about _____ inch(es) long.

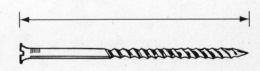

6. The washer is about _____ inch(es) wide.

Measuring with a Ruler

- The longest line halfway between the inch mark shows $\frac{1}{2}$ inch.

- The two next longest lines show $\frac{1}{4}$-inch increases.

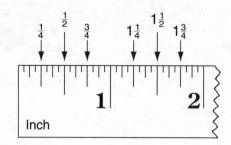

▶ Use a ruler to measure the length of each figure to the **nearest $\frac{1}{2}$ inch.**

1. _____ $4\frac{1}{2}$ inches

2. _____ inches

3. _____ inches

4. _____ inches

▶ Use a ruler to measure the length of each figure to the **nearest $\frac{1}{4}$ inch.**

5. _____ inches

6. _____ inches

7. _____ inches

8. _____ inch

Draw the Measurements

▶ Draw lines for the following lengths.

1. $4\frac{1}{2}$ inches

2. $1\frac{1}{2}$ inches

3. $3\frac{1}{2}$ inches

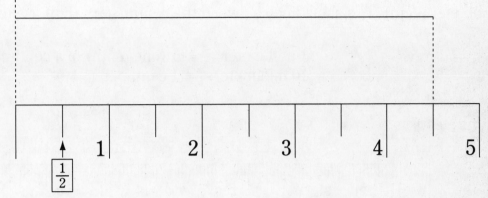

4. $3\frac{3}{4}$ inches

5. $4\frac{1}{4}$ inches

6. $2\frac{1}{2}$ inches

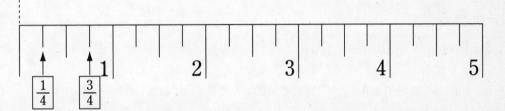

7. $3\frac{3}{4}$ inches

8. $1\frac{3}{8}$ inches

9. $4\frac{7}{8}$ inches

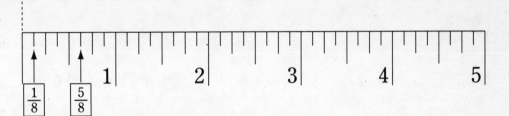

10. $4\frac{7}{16}$ inches

11. $2\frac{3}{4}$ inches

12. $4\frac{1}{4}$ inches

Measuring Inches and Centimeters

Measurements on a ruler are often shown with both the traditional and the metric system. The metric ruler shows centimeters and millimeters. The numbered lines are centimeters (cm). The short lines between the centimeters (cm) are millimeters (mm).

$$1 \text{ millimeter} = .1 \text{ (one-tenth) of a centimeter}$$
$$10 \text{ millimeters} = 1 \text{ centimeter}$$

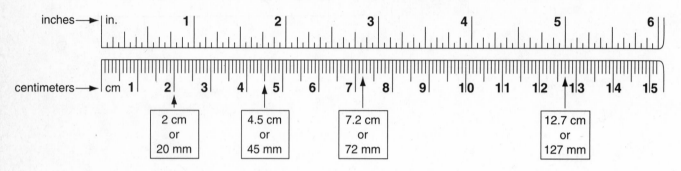

▶ Circle the best answer.

1. 1 inch is about   **a)** 2 cm   **b)** 2.5 cm   **c)** 30 mm

2. 4 inches is about   **a)** 98 mm   **b)** 10.5 cm   **c)** 10.2 cm

▶ Use the ruler to measure the length of each figure.

3. inches (in.) _____   **9.** inches (in.) _____

4. millimeters (mm) _____   **10.** millimeters (mm) _____

5. centimeters (cm) _____   **11.** centimeters (cm) _____

6. inches (in.) _____   **12.** inches (in.) _____

7. millimeters (mm) _____   **13.** millimeters (mm) _____

8. centimeters (cm) _____   **14.** centimeters (cm) _____

Traditional and Metric Measurements

Traditional Measurement

▶ Use a ruler to measure each wrench head in inches.

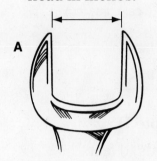 A

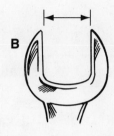 B

1. _____ in.   **2.** _____ in.

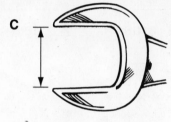 C

3. _____ in.

▶ Match a wrench from above with the nut of the same size. Write the letter of the wrench in the space provided.

4. _____
   letter

5. _____
   letter

6. _____
   letter

Metric Measurement

▶ Use a ruler to measure each wrench head in millimeters. (Use the ruler on page 16 if conversion from inches to millimeters is needed.)

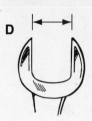 D

 E

7. _____ mm   **8.** _____ mm

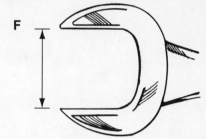 F

9. _____ mm

▶ Match a wrench from above with the nut of the same size. Write the letter of the wrench in the space provided.

10. _____
   letter

11. _____
   letter

12. _____
   letter

Converting Length Measurements

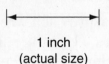 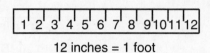 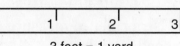

1 inch
(actual size)

12 inches = 1 foot
(scale drawing)

3 feet = 1 yard
(scale drawing)

Multiply to change from a larger unit to a smaller unit.

Example A

Change feet to inches.

larger unit		smaller unit

3 feet = ___?___ inches

12 in. = 1 ft.

$3 \times 12 = 36$

There are 36 inches in 3 feet.

Example B

Change yards to feet.

larger unit		smaller unit

4 yards = ___?___ feet

3 ft. = 1 yd.

$4 \times 3 = 12$

There are 12 feet in 4 yards.

▶ Change each measurement from larger units to smaller units. Remember, changing to a smaller unit means you will have more units.

1. 2 feet = __24__ inches

12 in. = 1 ft.

$2 \times 12 = 24$

2. 5 yards = _____ feet

3. 3 yards = _____ inches

$3 \times 12 = 36$

Find the number of inches in 1 yd.

4. 6 feet = _____ inches

5. 7 yards = _____ feet

6. 5 feet 8 inches = _____ inches

5×12 $60 + 8$

Multiply the feet by inches per foot. Then add the inches.

7. 3 feet 11 inches = _____ inches

8. 4 yards 2 feet = _____ feet

9. 6 yards 5 inches = _____ inches

6×36 $216 + 5$

Multiply yards by inches per yard. Add the inches.

10. 2 yards 1 foot 3 inches =
_____ inches

More Conversions

> 1 foot (ft.) = 12 inches (in.)
> 1 yard (yd.) = 3 ft. = 36 in.

> **Divide** to change from a smaller unit to a larger unit.

Example A
Change inches to feet.

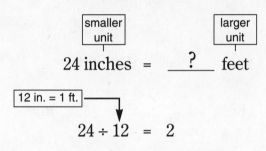

There are 24 inches in 2 feet.

Example B
Change feet to yards.

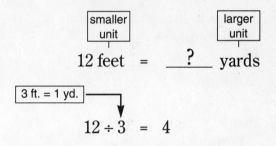

There are 12 feet in 4 yards.

▶ Change each measurement from smaller units to larger units. Remember, changing to a larger unit means you will have fewer units.

1. 48 inches = _____ feet

> 48 ÷ 12

Divide the inches.

2. 36 inches = _____ feet

3. 15 feet = _____ yards

4. 72 inches = _____ yards

5. 108 inches = _____ feet

6. 10 feet = _____ yards __1__ foot

> 10 ÷ 3

Divide the feet. One foot is left over.

7. 5 feet = _____ yard _____ feet

8. 40 inches = _____ feet _____ inches

9. 19 inches = _____ foot _____ inches

10. 13 feet = _____ yards _____ foot

Rename the Measurement

> 1 foot (ft.) = 12 inches (in.)
> 1 yard (yd.) = 3 ft. = 36 in.
> 5,280 ft. = 1 mile (mi.)

▶ Rename each measurement to its equivalent value. Circle whether you multiply or divide.

1. inches to feet
   a) multiply or divide

   b) 48 in. = _____ ft.

2. feet to yards
   a) multiply or divide

   b) 6 ft. = _____ yd.

3. miles to feet
   a) multiply or divide

   b) 2 mi. = _____ ft.

4. feet to inches
   a) multiply or divide

   b) 4 ft. = _____ in.

5. yards to feet
   a) multiply or divide

   b) 3 yd. = _____ ft.

6. inches to yards
   a) multiply or divide

   b) 180 in. = _____ yd.

▶ Rename each measurement.

7. 46 in. = _____ yd. _____ in.

8. 17 in. = _____ ft. _____ in.

9. 4 ft. 6 in. = _____ in.

10. 67 in. = _____ ft. _____ in.

11. 3 yd. 1 ft. = _____ in.

12. 18 ft. 4 in. = _____ in.

Comparing Measurements

Sometimes you will need to compare measurements. When comparing measurements, change the measurements so that they have the same units.

Example

Which is longer, 3 feet or 32 inches?

Think: Change the larger unit (feet) to inches.

1 foot = 12 inches

$$\boxed{3 \times 12}$$

so, 3 feet = 36 inches

$$\boxed{36 \text{ inches}}$$

Answer: 3 feet is longer than 32 inches.

▶ Remember to rename measurements in the same unit before comparing them. Write $>$ (greater than), $<$ (less than), or $=$ (equal to) in the circle.

1. 18 inches $\bigcirc<$ 2 feet
$2 \times 12 = 24$ in.

7. 3 feet 4 inches $\bigcirc$ 44 inches

2. 7 feet $\bigcirc$ 2 yards

8. 7 yards 1 foot $\bigcirc$ 20 feet

3. 3 yards $\bigcirc$ 110 inches

9. 2 yards 5 inches $\bigcirc$ 75 inches

4. 10 feet $\bigcirc$ 3 yards

10. 4 feet 4 inches $\bigcirc$ 60 inches

5. 37 inches $\bigcirc$ 3 feet

11. 24 feet 10 inches $\bigcirc$ 9 yards

6. 144 inches $\bigcirc$ 4 yards

12. 4 yards 2 feet $\bigcirc$ 14 feet

Adding Lengths

There are many times when you may want to add different lengths together. Sometimes it may be necessary to regroup.

Example A

Add the lengths. No regrouping is necessary.

$$4 \text{ ft. } 4 \text{ in. } + 2 \text{ ft. } 3 \text{ in.}$$

$$\begin{array}{r} 4 \text{ ft. } 4 \text{ in.} \\ + 2 \text{ ft. } 3 \text{ in.} \\ \hline 6 \text{ ft. } 7 \text{ in.} \end{array}$$

Example B

Add the lengths. Regrouping is necessary.

$$6 \text{ ft. } 7 \text{ in. } + 3 \text{ ft. } 8 \text{ in.}$$

$$\begin{array}{r} 6 \text{ ft. } 7 \text{ in.} \\ + 3 \text{ ft. } 8 \text{ in.} \\ \hline 9 \text{ ft. } 15 \text{ in.} \end{array} = 10 \text{ ft. } 3 \text{ in.}$$

simplified

↑ regroup
12 in. + 3 in.

▶ Add the measurements. Remember to regroup and simplify when necessary.

1. 2 ft. 5 in. + 5 ft. 9 in.

$$\begin{array}{r} 2 \text{ ft. } 5 \text{ in.} \\ + 5 \text{ ft. } 9 \text{ in.} \\ \hline 7 \text{ ft. } 14 \text{ in.} \end{array} = \underline{\hspace{1cm}} \text{ ft. } \underline{\hspace{1cm}} \text{ in.}$$

simplified

2. 5 ft. 7 in. + 3 ft. 3 in.

3. 3 yd. 5 ft. + 6 yd. 2 ft.
(Remember: 3 ft. = 1 yd.)

4. 1 ft. 8 in. + 7 ft. 3 in.

5. 10 yd. 2 ft. + 3 yd.

6. 4 ft. 9 in. + 15 in.

Subtracting Lengths

There are times when you will need to subtract different lengths. As in addition, sometimes you may need to regroup.

Example A

Subtract the numbers. No regrouping is necessary.

5 ft. 9 in. – 2 ft. 7 in.

$$\begin{array}{r} 5 \text{ ft. } 9 \text{ in.} \\ -\ 2 \text{ ft. } 7 \text{ in.} \\ \hline 3 \text{ ft. } 2 \text{ in.} \end{array}$$

Example B

Subtract the numbers. Regrouping is necessary.

10 yd. 1 ft. – 4 yd. 2 ft.

$$\begin{array}{r} \overset{9}{\cancel{10}} \text{ yd. } \overset{4}{\cancel{1}} \text{ ft.} \\ -\ 4 \text{ yd. } 2 \text{ ft.} \\ \hline 5 \text{ yd. } 2 \text{ ft.} \end{array}$$

regroup
10 yd. = 9 yd. 3 ft.
so
10 yd. 1 ft. = 9 yd. 4 ft.

▶ Subtract the measurements. Remember to regroup when necessary.

1. 4 ft. 7 in. – 2 ft. 9 in.

$$\begin{array}{r} \overset{3}{\cancel{4}} \text{ ft. } \overset{19}{\cancel{7}} \text{ in.} \\ -\ 2 \text{ ft. } 9 \text{ in.} \\ \hline \underline{} \text{ ft. } \underline{} \text{ in.} \end{array}$$

regroup
1 ft. = 12 in.

2. 3 yd. 2 ft. – 2 yd. 1 ft.

3. 7 ft. 8 in. – 6 ft. 7 in.

4. 2 ft. 10 in. – 1 ft. 11 in.

5. 5 yd. 2 ft. – 2 yd.

6. 10 yd. – 4 yd. 2 ft.

Review Your Skills

1. Write the measurements in order from smallest to largest.

70 inches   2 yards   3 feet

_____   _____   _____
smallest   largest

2. Circle the most reasonable answer.

The length of a shoe box is about:

a) 1 inch   **c)** 1 foot

b) 1 mile   **d)** 1 yard

3. Choose the object that is most reasonable for a measurement of 2 yards 6 inches.

a) sailboat **b)** train **c)** bicycle

4. Change each measurement to the given unit.

a) 2 feet = _____ inches

b) 9 feet = _____ yards

c) 36 inches = _____ feet

5. Use the length of 1 unit to estimate the length of the line segment.

The line segment is about _____ units long.

6. The arrow reads:

_____ feet _____ inches

7. Add the measurements. Regroup and simplify if necessary.

2 yd. 1 ft. + 1 yd. 2 ft.

8. Decide which measurement is longer and by how much.

15 inches or 1 foot

a) Which is longer? _____

b) How much longer is the larger measurement? _____

Using a Map Scale

A map is a reduced drawing of actual distances to fit on a piece of paper. The scale compares the actual distances to those drawn on a map.

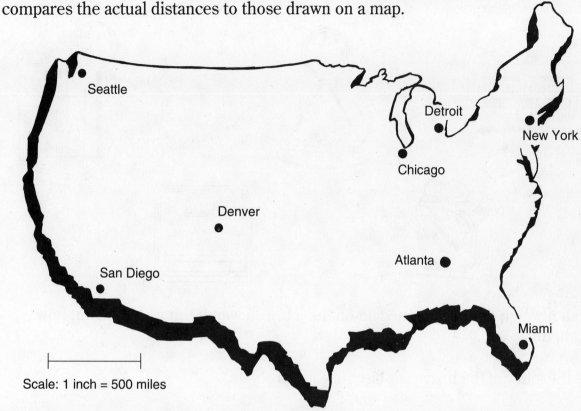

Scale: 1 inch = 500 miles

Each 1 inch on the map will equal about 500 miles of actual distance.

Example: About what is the distance from Denver to Seattle?

The measurement is a little over 2 inches ($2 \times 500 = 1,000$). So, a reasonable estimate is 1,200 miles. (Anywhere from 1,100 to 1,400 would be acceptable.)

▶ Use a ruler and the map scale to estimate the distances. All reasonable answers are acceptable.

1. Chicago to Detroit is
about _____ miles.

2. Denver to Atlanta is
about _____ miles.

3. Atlanta to Miami is
about _____ miles.

4. New York to Atlanta is
about _____ miles.

5. Seattle to San Diego is
about _____ miles.

6. San Diego to Miami is
about _____ miles.

Weight

There are many different kinds of scales that are used to weigh objects.

People often weigh things at home and on the job. You weigh things to find out how heavy something is.

▶ Circle the letter of the heaviest object in each problem.

1. a) baseball   **b)** Ping-Pong ball   **c)** bowling ball

2. a) bicycle   **b)** bus   **c)** car

3. a) coconut   **b)** apple   **c)** strawberry

▶ Circle the letter of the lightest object in each problem.

4. a) broom   **b)** vacuum cleaner   **c)** lawn mower

5. a) elephant   **b)** cat   **c)** mouse

6. a) grapefruit   **b)** watermelon   **c)** orange

Ounces, Pounds, and Tons

The ounce (oz.), pound (lb.), and ton (tn.) are units used to measure weight.

$$16 \text{ ounces (oz.)} = 1 \text{ pound (lb.)}$$
$$2,000 \text{ pounds} = 1 \text{ ton (tn.)}$$

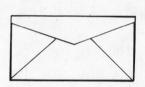  

about 1 ounce   about 1 pound   about 1 ton

▶ Would you use ounces, pounds, or tons to weigh these objects? Write your answers in the spaces provided.

1. A bag of potatoes _____   **4.** A watermelon _____

2. 5 pencils _____   **5.** A slice of cheese _____

3. A truck _____   **6.** A bowling ball _____

▶ Which unit of weight is most useful for measuring each of the following? Put the objects in the correct column below.

Adult person	Flashlight battery	Bag of groceries
Cement mixer	Trailer truck	Bar of soap
Sofa	Candy bar	Airplane

Ounces	Pounds	Tons

Reading Weight Scales

Study the weight scale.

Pounds can be shown by the term *lb.* or *lbs.*

This scale holds up to 4 pounds.

On this scale, each mark between 0 and 1 pound stands for 1 ounce. (16 ounces = 1 pound)

The arrow is pointing at 8 ounces.

▶ What do each of the following scales read? Write your answer in the space provided.

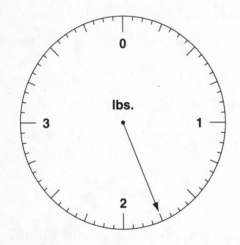

1. _____ pound _____ ounces

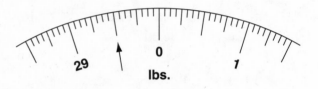

3. a) The scale holds up to _____ pounds.

   b) The arrow reads:

   _____ pounds _____ ounces

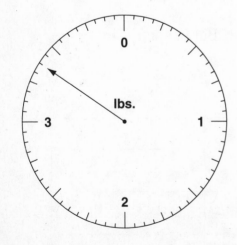

2. _____ pounds _____ ounces

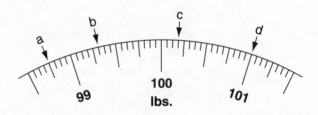

4. a) _____ pounds _____ ounces

   b) _____ pounds _____ ounces

   c) _____ pounds _____ ounces

   d) _____ pounds _____ ounces

Weights with Fractions

16 ounces = 1 pound

Sometimes weight scales are divided into sixteenths ($\frac{1}{16}$) of a pound. Parts out of 16 can be shown with fractions. An easy way to read this type of scale is to count the marks.

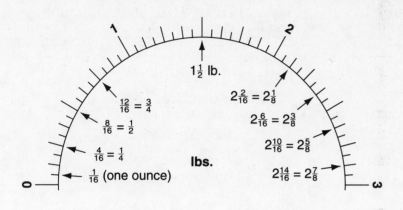

$1\frac{1}{2}$ lb.

$\frac{12}{16} = \frac{3}{4}$

$\frac{8}{16} = \frac{1}{2}$

$\frac{4}{16} = \frac{1}{4}$

lbs.

$\frac{1}{16}$ (one ounce)

$2\frac{2}{16} = 2\frac{1}{8}$

$2\frac{6}{16} = 2\frac{3}{8}$

$2\frac{10}{16} = 2\frac{5}{8}$

$2\frac{14}{16} = 2\frac{7}{8}$

- Study each scale carefully before reading it.

- Each mark between 0 and 1 stands for 1 ounce or $\frac{1}{16}$ of a pound.

▶ Find the marked values. Simplify all fractions.

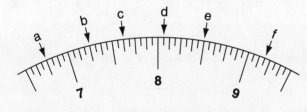

1. a) _____ lb.   **d)** _____ lb.

b) _____ lb.   **e)** _____ lb.

c) _____ lb.   **f)** _____ lb.

2. a) _____ lb.

b) _____ lb.

c) _____ lb.

d) _____ lb.

e) _____ lb.

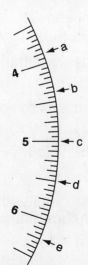

3. Label the arrows with the correct letters.

a) $3\frac{3}{16}$ lb.

b) $2\frac{3}{4}$ lb.

c) $4\frac{1}{16}$ lb.

d) $3\frac{1}{2}$ lb.

e) $2\frac{1}{4}$ lb.

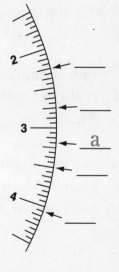

Converting Weight Measurements

16 ounces (oz.) = 1 pound (lb.)	2,000 pounds = 1 ton (tn.)

Sometimes you will want to change from one weight size to another by multiplying or dividing.

Multiply to change from a larger unit to a smaller unit.	**Divide** to change from a smaller unit to a larger unit.

Example A	**Example B**
Change pounds to ounces.	Change ounces to pounds.

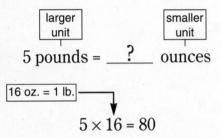

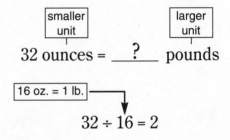

$5 \times 16 = 80$	$32 \div 16 = 2$
There are 80 ounces in 5 pounds.	There are 2 pounds in 32 ounces.

▶ Change the weights to equivalent measurements. Circle whether you must multiply or divide.

1. pounds to ounces
   a) multiply or divide
   b) 3 pounds = _____ ounces

2. ounces to pounds
   a) multiply or divide
   b) 48 ounces = _____ pounds

3. pounds to tons
   a) multiply or divide
   b) 6,000 pounds = _____ tons

4. pounds to ounces
   a) multiply or divide
   b) 10 pounds = _____ ounces

5. ounces to pounds
   a) multiply or divide
   b) 192 ounces = _____ pounds

6. pounds to ounces
   a) multiply or divide
   b) 15 pounds = _____ ounces

7. ounces to pounds
   a) multiply or divide
   b) 96 ounces = _____ pounds

8. tons to pounds
   a) multiply or divide
   b) 6 tons = _____ pounds

Rename the Measurements

> 16 ounces (oz.) = 1 pound (lb.)
>
> 2,000 pounds = 1 ton (tn.)

▶ Rename each measurement to its equivalent value.

1. 128 oz. = _____ lb.

2. 12 lb. = _____ oz.

3. 20,000 lb. = _____ tn.

4. 2 lb. 16 oz. = _____ lb.

5. $\frac{1}{2}$ lb. = _____ oz.

6. 20 oz. = _____ lb. _____ oz.

7. 11 lb. = _____ oz.

8. $4\frac{1}{2}$ lb. = _____ oz.

9. 90 oz. = _____ lb. _____ oz.

10. 4 lb. 3 oz. = _____ oz.

11. 4,150 lb. = _____ tn. _____ lb.

12. 72 oz. = _____ lb. _____ oz.

13. 40 oz. = _____ lb. _____ oz.

14. 1 tn. 450 lb. = _____ lb.

Estimating with Decimal and Fraction Weights

Sometimes it is necessary to compare decimal and fraction weights. They are easier to compare if you change them to the same unit.

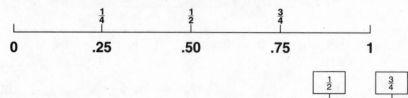

Example A: Is .58 closer to $\frac{1}{2}$ or $\frac{3}{4}$? **Think:** .58 is closer to .50 than .75 so .58 is closer to $\frac{1}{2}$.

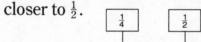

Example B: Is .31 closer to $\frac{1}{4}$ or $\frac{1}{2}$? **Think:** .31 is closer to .25 than .50 so .31 is closer to $\frac{1}{4}$.

▶ Circle the decimal weight that comes closest to the weight shown on the scale.

1.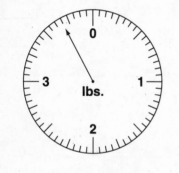

a) 3.25

b) 3.95

c) 3.49

d) 3.65

2.

a) 1.15

b) 1.75

c) 1.37

d) 1.59

▶ Circle the letter of your choice for each problem.

3. Jessie wants to buy $4\frac{1}{4}$ pounds of potato salad. Which label is closest to $4\frac{1}{4}$ pounds?

a) Net Weight 4.30 lbs.
b) Net Weight 4.44 lbs.
c) Net Weight 4.87 lbs.
d) Net Weight 4.14 lbs.

4. A recipe calls for $2\frac{1}{2}$ pounds of ground beef. Which label is closest to $2\frac{1}{2}$ pounds?

a) Net Weight 2.98 lbs.
b) Net Weight 2.45 lbs.
c) Net Weight 2.04 lbs.
d) Net Weight 2.64 lbs.

Adding and Subtracting Weights

To add or subtract units of weight, sometimes it may be necessary to regroup.

Example A
Add and regroup.
5 lb. 4 oz. + 7 lb. 20 oz.

$$\begin{array}{r} 5 \text{ lb.} \quad 4 \text{ oz.} \\ +\ 7 \text{ lb. } 20 \text{ oz.} \\ \hline 12 \text{ lb. } 24 \text{ oz.} = 13 \text{ lb. } 8 \text{ oz.} \end{array}$$

simplified

regroup to 1 lb. 8 oz.

Example B
Regroup and subtract.
5 lb. 3 oz. – 1 lb. 7 oz.

$$\begin{array}{r} {}^{4}\cancel{5} \text{ lb. } {}^{19}\cancel{3} \text{ oz.} \\ -\ 1 \text{ lb. } 7 \text{ oz.} \\ \hline 3 \text{ lb. } 12 \text{ oz.} \end{array}$$

regroup
5 lb. = 4 lb. 16 oz.
so 5 lb. 3 oz. = 4 lb. 19 oz.

▶ Add or subtract the weight measurements. Remember to regroup and simplify your answers when necessary.

1. 3 lb. 12 oz. + 6 lb. 9 oz.

$$\begin{array}{r} 3 \text{ lb. } 12 \text{ oz.} \\ +\ 6 \text{ lb. } \quad 9 \text{ oz.} \\ \hline 9 \text{ lb. } 21 \text{ oz.} = \underline{\quad} \text{ lb. } \underline{\quad} \text{ oz.} \end{array}$$

regroup   simplified

2. 9 lb. 5 oz. – 3 lb. 2 oz.

3. 8 tn. 800 lb. + 1 tn. 1,500 lb.

4. 6 lb. 5 oz. – 2 lb. 13 oz.

5. 4 lb. 14 oz. + 10 lb. 3 oz.

6. 7 lb. – 5 lb. 7 oz.

Review Your Skills

▶ Use your skills to answer the following problems.

John's weight in January

pounds

John's weight in April

pounds

1. About how many pounds did John gain? _____

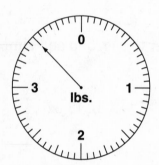

2. a) The smallest unit on the weight scale is _____ .

b) What does the scale read?

_____ pounds _____ ounces

3. Write in order from lightest to heaviest.

2 lb.   35 oz.   5 tn.

_____ _____ _____
lightest   heaviest

4. Complete:

a) 1 lb. 6 oz. = _____ oz.

b) 17 oz. = _____ lb. _____ oz.

5. Each box weighs about the same. Estimate the weight of one box.

_____ pounds

6. Is 4 pounds 13 ounces closer to 66 ounces or 5 pounds? _____

7. Decide which weight is heavier and by how much.

3 lb. 7 oz. or 64 oz.

a) Which is heavier? _____

b) How much heavier? _____

8. a) The scale holds up to _____ pounds.

b) The arrow points at:

_____ pounds _____ ounces

Shipping Costs

Postage Rate Table		
Weight not to exceed	Zone 1	Zone 2
1 lb.	$1.67	$1.92
2 lb.	1.69	2.18
3 lb.	1.84	2.43
4 lb.	1.98	2.61
5 lb.	2.08	2.76
6 lb.	2.15	2.88
7 lb.	2.23	3.00
8 lb.	2.29	3.23
9 lb.	2.43	3.45
10 lb.	2.55	3.68
11 lb.	2.70	3.93
12 lb.	2.85	4.15
13 lb.	2.97	4.38
14 lb.	3.11	4.60
15 lb.	3.23	4.83
16 lb.	3.37	5.07
17 lb.	3.51	5.29
18 lb.	3.63	5.52
19 lb.	3.78	5.74

"Weight not to exceed" means that if a package weighs more than the given number of pounds, you must round its weight to the next higher weight.

Example: How much would a package being shipped to Zone 1 cost if it weighs 7 pounds 1 ounce? 7 pounds 1 ounce must be rounded to 8 pounds. Find 8 pounds (lb.) on the Postage Rate Table. Then find where the 8 lb. row and the Zone 1 column meet. It would cost $2.29 to ship the package.

When mailing a package, you are charged according to zones. The zones shown above are for purchases mailed from Madison, Wisconsin. Zone 2 is farther away than Zone 1, so you must pay more when shipping to Zone 2.

▶ Use the map and Postage Rate Table to answer the questions.

	Item	Zone	Weight	Shipping Costs
1.	Blue jeans	1	2 lb. 2 oz.	
2.	Table lamp	2	18 lb.	
3.	Shorts	2	1 lb. 10 oz.	
4.	Answering machine	2	4 lb. 15 oz.	

Cups, Pints, Quarts, and Gallons

The cup, pint, quart, and gallon are standard units used to measure liquids.

1 cup (c.)

1 pint (pt.)
1 pint = 2 cups

1 quart (qt.)
1 quart = 2 pints

1 gallon (gal.)
1 gallon = 4 quarts

▶ Circle the most sensible liquid measurement. You can estimate the size of one object by thinking of another object whose size you already know.

1. About how much water does a bucket hold?

   a) 2 gallons   **b)** 2 quarts   **c)** 2 pints

2. About how much gas will fill an empty tank?

   a) 15 pints   **b)** 15 quarts   **c)** 15 gallons

3. Mr. Cody checked the oil in his car. About how much oil is needed to reach the full mark?

   a) 2 cups   **b)** 2 quarts   **c)** 2 gallons

4. About how much liquid will an average thermos hold?

   a) 1 cup   **b)** 1 quart   **c)** 5 gallons

5. About how much liquid does a large drinking glass hold?

   a) 1 pint   **b)** 1 quart   **c)** 1 gallon

Fluid Ounce

Fluid ounce is a standard unit of measurement that is used to measure liquids.

> ### A **fluid ounce** (fl. oz.) is 1 ounce of liquid measure.

2 tablespoons = 1 fluid ounce (1 fl. oz.)   8 fluid ounces = 1 cup

tablespoons

▶ Answer the questions using the measuring cup.

1. a) What fraction of the cup is filled with water? _____

   b) How many ounces is this? _____

2. a) What fraction of the cup is filled with water? _____

   b) Is this closer to 2 or 3 ounces?

_____

3. a) Shade the cup to show 6 ounces of orange juice.

   b) What fraction is this? _____

4. a) Shade the cup to show $\frac{1}{2}$ cup.

   b) How many ounces is this?

_____

Liter and Milliliter

The liter is the basic unit used to measure liquids in the metric system. The milliliter is used to measure small amounts of liquids.

1 liter (l) will fill about 4 glasses.

1 milliliter (ml) is about 2 drops.

1 liter (l) = 1,000 milliliters (ml)

▶ Estimate what each cup shows in milliliters (ml).

1.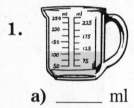

a) _____ ml   b) _____ ml

▶ Circle the letter of the measure you would use to find the amount of liquid for each.

2. Glass
   a) liters   **b)** milliliters

3. Spoon
   a) liters   **b)** milliliters

4. Bucket
   a) liters   **b)** milliliters

5. Medicine bottle
   a) liters   **b)** milliliters

6. Gas tank for a car
   a) liters   **b)** milliliters

▶ Shade the cups to estimate the milliliter measurements.

7.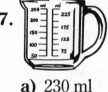

a) 230 ml   b) 75 ml

▶ Choose the most sensible measure, liters (l) or milliliters (ml).

8. A coffee cup will hold about

   150 _____.

9. An eyedropper will hold about

   5 _____.

10. A water pitcher will hold about

   2 _____.

11. A soup bowl holds about

   375 _____.

12. Three cans of paint total about

   12 _____.

Converting Liquid Measurements

2 cups (c.) = 1 pint (pt.)   2 pints (pt.) = 1 quart (qt.)   4 quarts (qt.) = 1 gallon (gal.)

Sometimes you will want to change from one liquid measurement size to another. You can do this by multiplying or dividing.

| **Multiply** to change from a larger unit to a smaller unit. | **Divide** to change from a smaller unit to a larger unit. |

Example A

Change gallons to quarts.

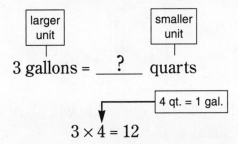

3 gallons = _____ quarts

$3 \times 4 = 12$

There are 12 quarts in 3 gallons.

Example B

Change quarts to gallons.

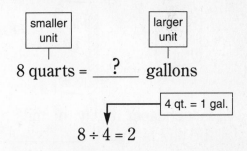

8 quarts = _____ gallons

$8 \div 4 = 2$

2 gallons contain 8 quarts.

▶ Circle whether you multiply or divide. Change to equivalent measures.

1. cups to pints
  a) multiply or (divide)
  b) 8 cups = _____ pints

2. quarts to pints
  a) multiply or divide
  b) 5 quarts = _____ pints

3. gallons to quarts
  a) multiply or divide
  b) 5 gallons = _____ quarts

4. pints to gallons
  a) multiply or divide
  b) 16 pints = _____ gallons

5. pints to quarts
  a) multiply or divide
  b) 6 pints = _____ quarts

6. pints to cups
  a) multiply or divide
  b) 7 pints = _____ cups

Real-Life Conversions

2 tablespoons (tbsp.) = 1 fluid ounce (fl. oz.)

1 cup (c.) = 8 fl. oz.

1 pint (pt.) = 2 cups = 16 fl. oz.

1 quart (qt.) = 2 pints = 32 fl. oz.

1 gallon (gal.) = 4 quarts = 128 fl. oz.

▶ Use the table above to answer the following questions.

 $.73    $1.49

1 gallon

1. Is it cheaper to buy the milk by the pint or quart? _____

4. 1 gallon of lemonade will fill how many cups? _____

2 tbsp. = 1 oz.

2. The recipe calls for 2 cups of water. How many fluid ounces is this? _____

5. How many tablespoons are in the bottle? _____

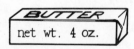

net wt. 4 oz.

3. A large recipe calls for 1½ cups of butter. How many sticks of butter is this? _____

6. The container holds 2 qt. 3.6 oz. If 1 serving equals 1 cup, how many servings does the container hold?

_____

Comparing Measurements

Sometimes you will need to compare liquid measurements. When comparing measurements, change each measurement to the same unit.

Example

Which is the greater amount, 3 gallons or 10 quarts?

Think: Change gallons to quarts.

$$1 \text{ gallon} = 4 \text{ quarts}$$

$$\boxed{3 \times 4}$$

so, 3 gallons = 12 quarts

$$\boxed{12 \text{ quarts}}$$

Answer: 3 gallons is a greater amount than 10 quarts.

▶ Remember to rename measurements in the same unit before comparing them. Write $>$ (greater than), $<$ (less than), or $=$ (equal to) in the circle.

1. 3 cups 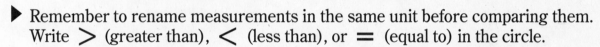 21 fluid ounces

7. 4 quarts 1 cup 1 gallon

2. 7 quarts ◯ 15 pints

8. 3 gallons 4 quarts 4 gallons

3. 6 pints ◯ 2 gallons

9. 2 pints 6 cups 9 cups

4. 3 cups ◯ 2 pints

10. 2 quarts 1 pint 6 pints

5. 2 gallons ◯ 8 quarts

11. 2 cups 20 fluid ounces

6. 12 quarts 5 gallons

12. 1 pint 1 cup 20 fluid ounces

Adding and Subtracting Liquid Measurements

To add or subtract liquid measurements, sometimes it may be necessary to regroup.

Example A

Add and regroup.
3 qt. 1 pt. + 2 qt. 1 pt.

$$
\begin{array}{r}
3 \text{ qt. 1 pt.} \\
+\ 2 \text{ qt. 1 pt.} \\
\hline
5 \text{ qt. 2 pt.} = 6 \text{ qt.}
\end{array}
$$

simplified

↑

| regroup to 1 quart |

Example B

Regroup and subtract.
4 gal. 2 qt. − 2 gal. 3 qt.

$$
\begin{array}{r}
^3 \quad\ ^6 \\
\cancel{4} \text{ gal. } \cancel{2} \text{ qt.} \\
-\ 2 \text{ gal. 3 qt.} \\
\hline
1 \text{ gal. 3 qt.}
\end{array}
$$

| regroup |
| 4 gal. = 3 gal. 4 qt. |
| so, 4 gal. 2 qt. = 3 gal. 6 qt. |

▶ Add or subtract the liquid measurements. Remember to regroup and simplify your answers when necessary.

1. 2 gal. 1 qt. + 3 gal. 3 qt.

$$
\begin{array}{r}
2 \text{ gal. 1 qt.} \\
+\ 3 \text{ gal. 3 qt.} \\
\hline
5 \text{ gal. 4 qt.} = \underline{\qquad} \text{ gal.}
\end{array}
$$

↑

regroup

4. 5 gal. 2 qt. − 4 gal. 3 qt.

2. 5 qt. 1 pt. − 1 qt. 2 pt.

5. 7 qt. 4 pt. + 2 qt. 1 pt.

3. 3 c. 3 fl. oz. + 1 c. 7 fl. oz.

6. 3 gal. 2 qt. − 2 gal. 3 qt.

Review Your Skills

▶ Use your skills to answer the following questions.

1. Write the units in order from smallest to largest.

a) qt., pt., c., gal., fl. oz.

_____ _____ _____ _____ _____
smallest   largest

b) l, ml

_____   _____
smaller   larger

2. 1 gallon is equal to:

a) _____ quarts

b) _____ pints

c) _____ cups

3. Would a cup, quart, or gallon be the most sensible unit of measure?

a) The amount of gasoline in a car _____

b) A glass of milk _____

4. Which one of the following is a unit of liquid measure? _____

a) pound

b) centimeter

c) liter

d) foot

$.88 $3.59

5. Is it cheaper to buy the milk by the quart or gallon? _____

6. Complete the following problems.

a) 3 qt. = _____ pt.

b) 3 gal. 2 qt. = _____ qt.

c) 8 qt. = _____ gal.

7. Circle the letter of the largest measurement.

a) 5 gal. **b)** 3 gal. 9 qt. **c)** 15 c.

8. Decide which measurement is larger and by how much.

5 pints or 4 quarts

a) Which is larger? _____

b) By how much? _____

Measuring Ingredients in Recipes

Measuring cups and spoons are used to measure ingredients for recipes.

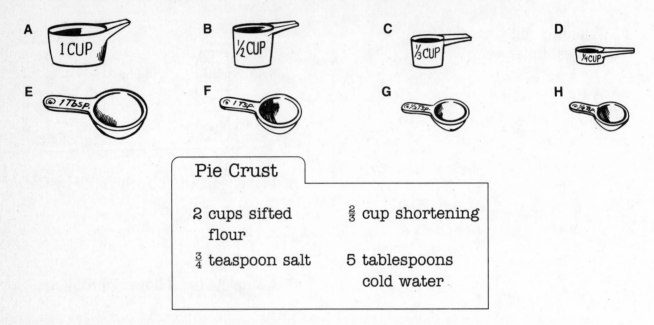

Pie Crust

2 cups sifted flour	$\frac{2}{3}$ cup shortening
$\frac{3}{4}$ teaspoon salt	5 tablespoons cold water

▶ Use the measuring cups and spoons to describe how you would measure the above ingredients to make a pie crust.

1. 2 cups sifted flour ___Fill up cup A twice.___

2. $\frac{3}{4}$ teaspoon salt _____

3. $\frac{2}{3}$ cup shortening _____

4. 5 tablespoons cold water _____

Cherry Pie Filling

$1\frac{1}{4}$ cups sugar	$\frac{1}{4}$ teaspoon salt
2 tablespoons flour	1 quart tart red cherries

▶ Describe how you would measure the ingredients to make a filling for a cherry pie.

5. $1\frac{1}{4}$ cups sugar _____

6. 2 tablespoons flour _____

7. $\frac{1}{4}$ teaspoon salt _____

8. 1 quart tart red cherries _____

Real-Life Applications

1. Name 6 things you can measure in a measuring cup.

   a) _____ **d)** _____

   b) _____ **e)** _____

   c) _____ **f)** _____

2. Nina's recipe called for 1 cup 3 fluid ounces of milk. How many ounces of milk did she need?

   _____ fluid ounces

3. Mom's recipe calls for $\frac{1}{4}$ cup of water for each pint of chili. How many cups of water will be needed for 6 quarts?

   _____ cups

4. Aretha plans to invite 8 friends to her birthday party. She wants to give each person 8 fl. oz. of orange punch. How many gallons will she need?

   _____ gallon(s)

5. Jesse serves soft drinks in 8-oz. cups. How many servings can he get from 2 gallons?

   _____ servings

Marshmallow Sauce

   1 cup sugar

   $\frac{1}{2}$ cup water

   16 marshmallows

   2 egg whites

---

6. Using the above recipe, how much water would you need to make:

   a) $\frac{1}{2}$ the recipe? _____ cup(s)

   b) twice the recipe? _____ cup(s)

   c) 3 times the recipe? _____ cup(s)

7. Ginger charges $.65 for an 8-oz. glass of soda. If it costs her $1.25 for each gallon, how much profit does she make per gallon?

   $_____

8. A recipe calls for $\frac{1}{2}$ cup of water for each quart of tomatoes. How many cups of water will be needed for 1 gallon?

   _____ cups

9. A recipe calls for 4 fl. oz. of cooking oil. How many tablespoons is this?

   _____ tablespoons

Temperature

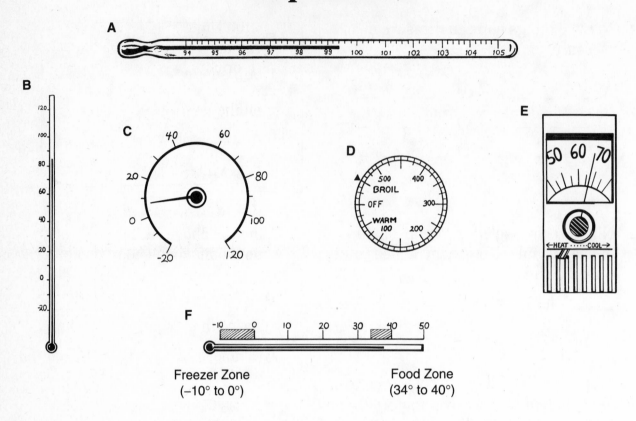

A

B

C

D

E

F

Freezer Zone
(−10° to 0°)

Food Zone
(34° to 40°)

▶ Write the letter of the scale you would use to measure the temperature. Then read each scale to estimate the temperature in degrees Fahrenheit (°F).

1. The temperature on a cold winter day   **a)** _____   **b)** _____ °F
                                                     letter   estimate

2. The temperature inside a refrigerator   **a)** _____   **b)** _____ °F
                                                       letter   estimate

3. The furnace setting on a thermostat   **a)** _____   **b)** _____ °F
                                                       letter   estimate

4. The temperature at the beach on a warm summer day   **a)** _____   **b)** _____ °F
                                                       letter   estimate

5. The temperature to cook a steak   **a)** _____   **b)** _____ °F
                                                       letter   estimate

6. The temperature of a sick person with a low fever   **a)** _____   **b)** _____ °F
                                                       letter   estimate

Fahrenheit Temperatures

The most common unit for measuring temperature in the United States is degrees Fahrenheit (°F). On the Fahrenheit scale, water freezes at 32° and boils at 212°. A Fahrenheit temperature can be written two ways: 68°F or 68°.

▶ Circle the letter of the most sensible temperature on the Fahrenheit thermometer.

1. Comfortable room temperature

    a) 10°F   **b)** 70°F   **c)** 90°F

2. Frozen ice cream

    a) 20°F   **b)** 35°F   **c)** 48°F

3. Very warm sunny day

    a) 30°F   **b)** 62°F   **c)** 89°F

4. A puddle of water starting to turn to ice

    a) 32°F   **b)** 45°F   **c)** 70°F

5. Very cold winter day

    a) 54°F   **b)** 40°F   **c)** 45°F   **d)** 12°F

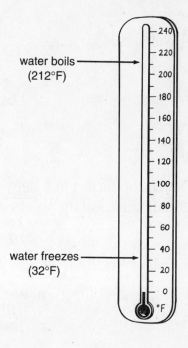

Fahrenheit

water boils (212°F)

water freezes (32°F)

▶ Complete the table of temperature changes across the country.

	Early Morning Temperature	Increase or Decrease	Evening Temperature	Above or Below Freezing
6.	45°F	Increase of 17°	62 °F	Above
7.	36°F	Decrease of 9°	_____ °F	_____
8.	15°F	Increase of 16°	_____ °F	_____
9.	28°F	Increase of 5°	_____ °F	_____
10.	80°F	Decrease of 7°	_____ °F	_____

Compare Fahrenheit and Celsius Temperatures

Traditional units of temperature are degrees Fahrenheit (F). Metric units of temperature are degrees Celsius (C), or centigrade. Most countries measure temperature in degrees Celsius.

▶ Use the temperature scales to answer the questions.

1. a) Water freezes at _____°C and _____°F.

   b) Water boils at _____°C and _____°F.

2. What is the difference between the freezing and boiling points of water on the

   a) Celsius thermometer? _____°C

   b) Fahrenheit thermometer? _____°F

3. If the indoor temperature is 20° Celsius, what is the reading in Fahrenheit degrees? _____°F

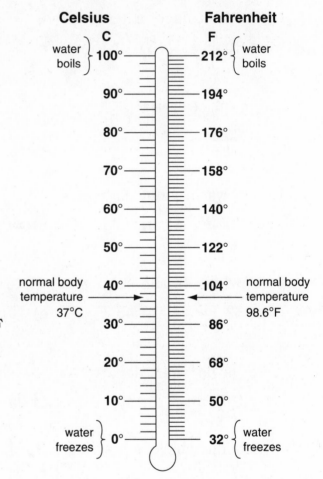

Celsius   **Fahrenheit**

water boils — 100°   212° — water boils

90° — 194°
80° — 176°
70° — 158°
60° — 140°
50° — 122°

normal body temperature 37°C → 40° — 104° ← normal body temperature 98.6°F

30° — 86°
20° — 68°
10° — 50°

water freezes — 0°   32° — water freezes

▶ Use the Celsius and Fahrenheit scales to compare each of the following. Estimate your answers.

4. 88°F is about _____°C.

5. 98°F is about _____°C.

6. 48°C is about _____°F.

7. 28°C is about _____°F.

▶ Circle the most sensible answer.

8. A comfortable temperature if you wanted to go swimming

   a) 36°F   b) 29°C   c) 75°C

9. Comfortable room temperature

   a) 92°F   b) 70°F   c) 36°C

10. An outdoor temperature that requires wearing a coat

    a) 4°C   b) 38°C   c) 78°F

Temperatures Below Zero

We use a negative sign to show temperatures below zero. The temperatures below are in degrees Fahrenheit.

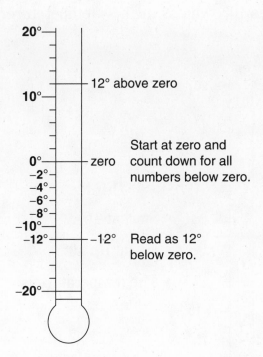

A temperature below zero will be colder than a temperature above zero.

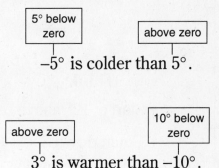

−5° is colder than 5°.

3° is warmer than −10°.

▶ Shade the thermometers to show the given temperature.

1. −10°	**2.** −4°	**3.** −30°	**4.** −40°

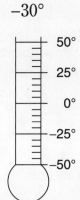
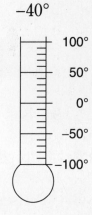

▶ Circle the warmer temperature.

5. a) −13° or −5°

   b) 16° or 18°

   c) −2° or 5°

   d) 7° or −7°

▶ Circle the colder temperature.

6. a) 8° or 10°

   b) 3° or −2°

   c) −1° or −4°

   d) −12° or 12°

Rising and Falling Temperatures

Falling Temperatures

If the temperature is 12° above zero and **drops** 20°, what is the new temperature?

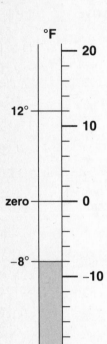

Think:
- Start at 12° above zero.
- Count **down** 20°.
- The temperature is –8°.

Rising Temperatures

If the temperature is 6° below zero and **rises** 10°, what is the new temperature?

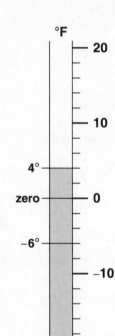

Think:
- Start at –6°.
- Count **up** 10°.
- The temperature is 4°.

▶ Find the **change** in temperature.

1. 9° rises to 20° $\underline{11°}$

2. 5° drops to –2° _____

3. –3° rises to 10° _____

4. –4° drops to –15° _____

5. 13° drops to –5° _____

6. 24° rises to 43° _____

▶ Find the **final** temperature.

7. 7°, rises 34° $\underline{41°}$

8. 8°, drops 5° _____

9. –12°, rises 17° _____

10. –2°, drops 11° _____

11. 5°, drops 5° _____

12. 65°, rises 23° _____

Health Thermometer

We use health thermometers to take people's temperatures. Thermometers like the one below use mercury and a scale to show temperature.

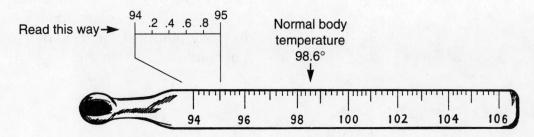

- Health thermometers use decimal places to show temperature.

- There are five equal parts between 94° and 95°, so each small mark stands for .2 (two-tenths) of a degree.

- Normal body temperature is 98.6 degrees Fahrenheit.

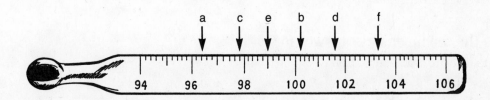

1. What does each of the marked temperatures read?

a) _____   **c)** _____   **e)** _____

b) _____   **d)** _____   **f)** _____

2. How much **above** normal body temperature is:

a) 100.8°? _2.2°_   **c)** 101.4°? _____

b) 99.2°? _____   **d)** 102.0°? _____

3. How much **below** normal body temperature is:

a) 98.2°? _.4°_   **c)** 97.4°? _____

b) 96.8°? _____   **d)** 95.0°? _____

Review Your Skills

▶ Use your skills to answer the following questions.

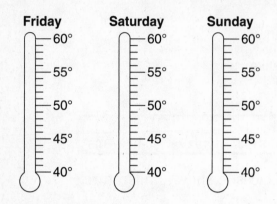

1. Shade the thermometer readings to show:

    a) 59° for Friday

    b) 44° for Saturday

    c) 52° for Sunday

2. Circle the most sensible answer for a comfortable summer day.

    a) 40° F   **b)** 98° F   **c)** 75° F

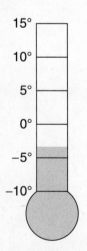

3. The thermometer shows an early morning temperature. Estimate the temperature reading. _____

4. How many degrees of temperature are between the boiling and freezing points of water on the Fahrenheit thermometer? _____

5. The weather forecast is for a high of 15° F and a low of −7° F. What is the difference in the high and low temperatures? _____

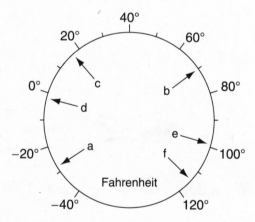

6. Estimate the marked temperature values:

    a) _____   **d)** _____

    b) _____   **e)** _____

    c) _____   **f)** _____

7. The temperature reading was 17° F at noon. At 6:00 P.M. it had dropped 20°. What was the temperature at 6:00 P.M.? _____

Windchill Table

Cold and wind make the air feel colder than the actual temperature. Windchill is an estimate of how cold the wind makes the air feel.

Windchill Table (°F)						
Wind Speed (mph)	Actual Temperature					
	50°	40°	30°	20°	10°	0°
5	48°	37°	27°	16°	7°	−5°
10	40°	28°	16°	3°	−9°	−22°
15	36°	22°	9°	−5°	−18°	−31°
20	32°	18°	4°	−10°	−24°	−39°
25	30°	16°	1°	−15°	−29°	−44°
30	28°	13°	−2°	−18°	−33°	−49°
35	27°	11°	−4°	−20°	−35°	−52°
40	26°	10°	−5°	−21°	−37°	−53°

Source: National Weather Service, NOAA, U.S. Commerce Department.

Example: What is the windchill when the wind speed is 20 mph and the actual temperature is 30°?

Step 1

Find the column that lists wind speed and find 20 mph.

Step 2

Find the place where the 20 mph wind speed row and the 30° temperature column meet.

For a 20 mph wind and a temperature of 30°, the windchill is 4°F.

▶ Use the windchill table to answer the following questions.

1. What is the windchill when:
   a) the wind is 15 mph and the actual temperature is 10°? _____
   b) the wind is 10 mph and the actual temperature is 50°? _____
   c) the wind is 35 mph and the actual temperature is 0°? _____

2. What is the difference in windchill temperatures between problems a and b? _____

3. What is the difference in windchill temperatures between problems b and c? _____

Time

Time can be measured in different units.

▶ Circle the most sensible answer.

1. What time of the day will the sun set?

   a) noon   **b)** morning   **c)** evening

2. About how long will it take Jeff to hit the water?

   a) 1 second   **b)** 1 minute   **c)** 1 hour

3. About how long does it take to walk one mile?

   a) 15 years   **b)** 15 hours   **c)** 15 minutes

Danville

55 miles   Mill Run

4. About how long will it take to travel by car from Danville to Mill Run?

   a) 10 minutes   **b)** 1 hour   **c)** 5 hours

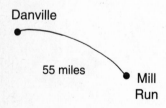

5. Which is the longest unit of time?

   a) week   **b)** month   **c)** year

Seconds, Minutes, and Hours

The second, minute, and hour are standard units that measure time.

60 seconds (sec.)	=	1 minute
60 minutes (min.)	=	1 hour
24 hours (hr.)	=	1 day

1 second

the average time it takes to say "one thousand one"

1 minute

the length of an average TV commercial

1 hour

the average time it takes you to walk 4 miles

▶ Choose either seconds, minutes, or hours to time the following:

1. Full-length movie _____

2. Cracking an egg _____

3. Driving to work _____

4. Eating breakfast _____

5. Coffee break _____

6. Sleeping _____

7. Taking out the trash _____

8. Noon to midnight _____

9. Working _____

10. Your heart beating 10 times _____

▶ Estimate the amount of time it would take you to do each of the following activities.

11. Go out for dinner _____

12. Wash the dishes _____

13. Eat lunch _____

14. Run a 100-yard dash _____

15. Watch a basketball game _____

16. Wash your hair _____

17. Boil water _____

18. Read the newspaper _____

19. Fall asleep _____

20. Do the grocery shopping _____

Digital Clocks

Digital clocks display time using digits and a colon. Digital times are usually read as they appear — 6:48.

The time on an analog clock can be read as:

- 48 minutes after 6
- 12 minutes to 7
- 6:48

▶ Use the digital clocks to complete the three different ways the time can be read.

1. a) _____ minutes after _____

   b) _____ minutes to _____

   c) _____ : _____

2. a) _____ minutes after _____

   b) _____ minutes to _____

   c) _____ : _____

3. a) _____ minutes after _____

   b) _____ minutes to _____

   c) _____ : _____

4. a) _____ minutes after _____

   b) _____ minute to _____

   c) _____ : _____

5. a) _____ minutes after _____

   b) _____ minutes to _____

   c) _____ : _____

6. When the time is 30 minutes or less past the hour, it is **not** necessary to read the time before the next hour.

   a) _____ minutes after _____

   b) _____ : _____

A.M. and P.M.

Noon
Midnight

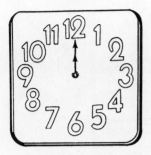

1 day = 24 hours

A.M. stands for the time from midnight to noon.
12:00 A.M. is midnight.

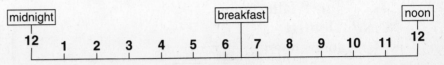

P.M. stands for the time from noon to midnight.
12:00 P.M. is noon.

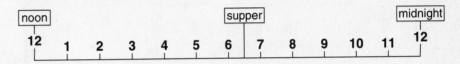

▶ Write the letters *A.M.* or *P.M.* for each time.

1. 3 o'clock in the afternoon _____

2. 8 o'clock in the morning _____

3. 1 hour before midnight _____

4. 4:25 in the afternoon _____

5. noon _____

6. sunrise _____

▶ Write the time shown on the clock, and then circle A.M. or P.M. to show what part of the day the activity belongs in.

7.

Raking leaves in the evening

_____ : _____ | A.M. or P.M. |

8.

Morning coffee break

_____ : _____ | A.M. or P.M. |

9.

Eating lunch

_____ :_____ | A.M. or P.M. |

10.

Time for a midnight snack

_____ : _____ | A.M. or P.M. |

Weekly Appointments

▶ Draw the hands on the clock and complete the given times. Be sure to circle A.M. or P.M. in each problem.

1. **MONDAY** July 9 (afternoon)
Dentist
four fifteen   quarter after

   a) __15__ minutes after __4__

   b) _____ : _____ A.M. or P.M.

2. **TUESDAY** July 10 (morning)
Grocery Shopping
eight forty-five   quarter to

   a) _____ minutes to _____

   b) _____ : _____ A.M. or P.M.

3. **WEDNESDAY** July 11 (afternoon)
Luncheon
twelve thirty

   a) _____ minutes after _____

   b) _____ : _____ A.M. or P.M.

4. **THURSDAY** July 12 (morning)
Baby-sitting
nine thirty-five

   a) _____ minutes to _____

   b) _____ : _____ A.M. or P.M.

5. **FRIDAY** July 13 (evening)
Birthday Party
five minutes past seven

   a) _____ minutes after _____

   b) _____ : _____ A.M. or P.M.

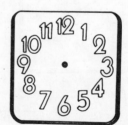

Elapsed Time

To find the time before or after a given time:

- Count the hours.
- Count the minutes.

Example A

What time is 2 hours 30 minutes before 4:00 P.M.?

Step 1	Step 2
Count the hours.	Count the minutes.
2 hours before 4:00 P.M. is 2:00 P.M.	30 minutes before 2:00 P.M. is 1:30 P.M.

2 hours 30 minutes before 4:00 P.M. is 1:30 P.M.

Example B

What time is 1 hour 15 minutes after 7:30 P.M.?

Step 1	Step 2
Count the hours.	Count the minutes.
1 hour after 7:30 P.M. is 8:30 P.M.	15 minutes after 8:30 P.M. is 8:45 P.M.

1 hour 15 minutes after 7:30 P.M. is 8:45 P.M.

▶ Find the times that are asked for in the problems below.

1. What time is 3 hours 15 minutes before 11:00 A.M.? _____

2. What time is 2 hours 30 minutes after 5:00 P.M.? _____

3. What time is 5 hours 10 minutes before 10:00 A.M.? _____

4. What time is 1 hour 30 minutes after 6:30 P.M.? _____

5. What time is 4 hours 15 minutes after 5:25 A.M.? _____

6. What time is 6 hours 20 minutes before 9:15 P.M.? _____

7. What time is 45 minutes after 5:45 P.M.? _____

8. What time is 35 minutes before 12:05 P.M.? _____

Count the Hours and Minutes Worked

Sometimes you need to find how much time you worked in a day.

Example: How many hours and minutes did Sherry work if she started at 8:30 A.M. and worked until 2:45 P.M.?

Started work at
8:30 A.M.

Finished work at
2:45 P.M.

<u>Step 1:</u> Count the hours. There are 6 hours between 8:30 A.M. and 2:30 P.M.

<u>Step 2:</u> Count the minutes. There are 15 minutes between 2:30 P.M. and 2:45 P.M.

Sherry worked 6 hours and 15 minutes.

▶ Count the hours and minutes worked.

1. Started work at
7:00 A.M.

Finished work at
3:30 P.M.

_____ hours _____ minutes

3. Started work at
8:00 A.M.

Finished work at
4:30 P.M.

_____ hours _____ minutes

2. Started work at
10:30 A.M.

Finished work at
2:15 P.M.

_____ hours _____ minutes

4. Started work at
8:30 A.M.

Finished work at
3:20 P.M.

_____ hours _____ minutes

Time Zones

The earth's rotation causes the sun to rise first in the east. It is daylight in New York City 3 hours before it is daylight in Los Angeles. To allow for this difference in daylight, we have time zones.

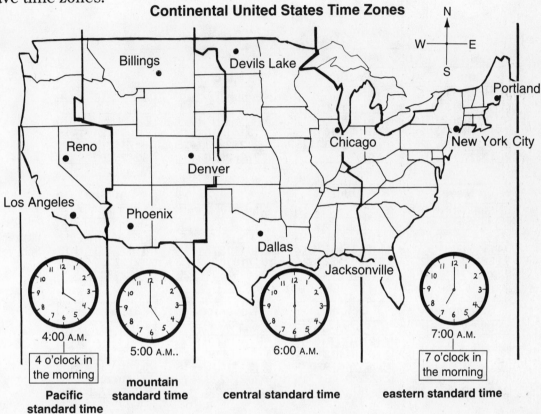

Continental United States Time Zones

4:00 A.M.
4 o'clock in the morning
Pacific standard time

5:00 A.M.
mountain standard time

6:00 A.M.
central standard time

7:00 A.M.
7 o'clock in the morning
eastern standard time

▶ Use the map and time zones to answer the questions.

1. When it is 7 o'clock in the morning in New York City, what time is it in:

    a) Chicago? 6:00 A.M.   **b)** Denver? _____   **c)** Los Angeles? _____

2. When it is 5 o'clock in the afternoon in Dallas, what time is it in:

    a) Reno? _____   **c)** Devils Lake? _____   **e)** Phoenix? _____

    b) Billings? _____   **d)** Jacksonville? _____   **f)** Portland? _____

3. Name the time zone.

    a) At 12:00 midnight Pacific standard time, it is 3:00 A.M. _____ standard time.

    b) At 2:30 P.M. Pacific standard time, it is 3:30 P.M. _____ standard time.

4. Name the time.

    a) At 8:00 P.M. Pacific standard time, it is _____ central standard time.

    b) At _____ eastern standard time, it is 1:00 P.M. Pacific standard time.

Using a Calendar

We use a calendar to keep track of months, weeks, and days.

1 week = 7 days	1 year = 12 months = 52 weeks = 365 days

Each week → starts with Sunday.

January

S	M	T	W	T	F	S
1	2	3	4	5	6	7
8	9	10	11	12	13	14
15	16	17	18	19	20	21
22	23	24	25	26	27	28
29	30	31				

February

S	M	T	W	T	F	S
				1	2	3	4
5	6	7	8	9	10	11
12	13	14	15	16	17	18
19	20	21	22	23	24	25
26	27	28				

March

S	M	T	W	T	F	S
			1	2	3	4
5	6	7	8	9	10	11
12	13	14	15	16	17	18
19	20	21	22	23	24	25
26	27	28	29	30	31	

April

S	M	T	W	T	F	S
						1
2	3	4	5	6	7	8
9	10	11	12	13	14	15
16	17	18	19	20	21	22
23	24	25	26	27	28	29
30						

May

S	M	T	W	T	F	S
	1	2	3	4	5	6
7	8	9	10	11	12	13
14	15	16	17	18	19	20
21	22	23	24	25	26	27
28	29	30	31			

June

S	M	T	W	T	F	S
				1	2	3
4	5	6	7	8	9	10
11	12	13	14	15	16	17
18	19	20	21	22	23	24
25	26	27	28	29	30	

July

S	M	T	W	T	F	S
						1
2	3	4	5	6	7	8
9	10	11	12	13	14	15
16	17	18	19	20	21	22
23	24	25	26	27	28	29
30	31					

August

S	M	T	W	T	F	S
		1	2	3	4	5
6	7	8	9	10	11	12
13	14	15	16	17	18	19
20	21	22	23	24	25	26
27	28	29	30	31		

The same date may fall on a different day from year to year.

September

S	M	T	W	T	F	S
					1	2
3	4	5	6	7	8	9
10	11	12	13	14	15	16
17	18	19	20	21	22	23
24	25	26	27	28	29	30

October

S	M	T	W	T	F	S
1	2	3	4	5	6	7
8	9	10	11	12	13	14
15	16	17	18	19	20	21
22	23	24	25	26	27	28
29	30	31				

November

S	M	T	W	T	F	S
			1	2	3	4
5	6	7	8	9	10	11
12	13	14	15	16	17	18
19	20	21	22	23	24	25
26	27	28	29	30		

December

S	M	T	W	T	F	S
					1	2
3	4	5	6	7	8	9
10	11	12	13	14	15	16
17	18	19	20	21	22	23
24	25	26	27	28	29	30
31						

▶ Use the calendar above to answer the questions.

1. How many days are in May? _____

2. How many Sundays are in October? _____

3. What day of the week is July 4? _____

4. What day is the last Friday in February? _____

5. What day of the week is 8 days after August 7? _____

6. How many months are there from April 1 to June 1? _____

7. Circle the days for the start of each season on the calendar.

   a) spring — March 21
   b) summer — June 21
   c) autumn (fall) — September 23
   d) winter — December 21

8. Kevin works Monday through Friday.

   a) How many working days does he have in February? _____

   b) In what months will Kevin have 23 working days? _____

_____

Reading and Writing Dates

We often use numbers to write dates. For example, February 14 can be written as 2/14. The months are numbered in order from 1 to 12 starting with January.

Number	Month	Number	Month
1	January	7	July
2	February	8	August
3	March	9	September
4	April	10	October
5	May	11	November
6	June	12	December

- The first number on the watch displays the month. The ninth month is September.

- The second number displays the day. The tenth day of the month is shown here.

▶ Write the month and day for each display.

1. a) [3   7] _____ _____
month day   month   day

c) [8   3] _____ _____
month   day

b) [11   21] _____ _____
month   day

d) [5   30] _____ _____
month   day

Dates are written with the month first, then the day of the month followed by the last two digits of the year. The slash mark **(/)** separates the month, day, and year.

month   day   year

Example: 4/15/91   Date: April 15, 1991

▶ Use numbers to write the dates.

2. a) October 5, 1962   Date: _____

b) September 6, 1995   Date: _____

c) July 31, 1984   Date: _____

d) December 25, 1999   Date: _____

Converting Time

60 seconds (sec.)	=	1 minute (min.)	
60 minutes	=	1 hour (hr.)	
24 hours	=	1 day (d.)	
7 days	=	1 week (wk.)	

365 days	=	1 year (yr.)	
12 months (mo.)	=	1 year	
52 weeks	=	1 year	
100 years	=	1 century (cent.)	

Multiply to change from a larger unit to a smaller unit.

Divide to change from a smaller unit to a larger unit.

Example A

Change the number of weeks to days.

larger unit   smaller unit

4 weeks = ___?___ days

7 d. = 1 wk.

$4 \times 7 = 28$

There are 28 days in 4 weeks.

Example B

Change the number of months to years.

smaller unit   larger unit

36 months = ___?___ years

12 mo. = 1 yr.

$36 \div 12 = 3$

There are 36 months in 3 years.

▶ Circle whether you should multiply or divide. Then change each amount to the unit of measure being asked for.

1. years to months

   a) multiply or divide

   b) 3 yr. = _____ mo.

2. hours to days

   a) multiply or divide

   b) 48 hr. = _____ d.

3. weeks to days

   a) multiply or divide

   b) 8 wk. = _____ d.

4. minutes to hours

   a) multiply or divide

   b) 240 min. = _____ hr.

▶ Change each amount to the unit being asked for.

5. 4 hr. 16 min. = _____ min.

6. 4 yr. 6 mo. = _____ mo.

7. 1 yr. 24 wk. = _____ wk.

8. 2 d. 10 hr. = _____ hr.

Adding and Subtracting Time

To add or subtract units of time, it may be necessary to regroup.

Example A	Example B
Add and regroup.	Regroup and subtract.
3 hr. 25 min. + 5 hr. 48 min.	4 min. 36 sec. – 2 min. 50 sec.

Example A:

$$3 \text{ hr. } 25 \text{ min.}$$
$$+ 5 \text{ hr. } 48 \text{ min.}$$
$$\overline{8 \text{ hr. } 73 \text{ min.}} = 9 \text{ hr. } 13 \text{ min.}$$
simplified

regroup to 1 hr. 13 min.

Example B:

         3   96
$$\cancel{4} \text{ min. } \cancel{36} \text{ sec.}$$
$$- 2 \text{ min. } 50 \text{ sec.}$$
$$\overline{1 \text{ min. } 46 \text{ sec.}}$$

regroup
4 min. = 3 min. 60 sec. so
4 min. 36 sec. = 3 min. 96 sec.

▶ Add or subtract the time measurements. Remember to regroup and simplify your answers when necessary.

1. 2 hr. 20 min. + 5 hr. 50 min.

$$2 \text{ hr. } 20 \text{ min.}$$
$$+ 5 \text{ hr. } 50 \text{ min.}$$
$$\overline{7 \text{ hr. } 70 \text{ min.}} = \underline{\quad} \text{ hr. } \underline{\quad} \text{ min.}$$
simplified

regroup

4. 8 hr. 15 min. – 2 hr. 30 min.

2. 8 min. 15 sec. – 3 min. 58 sec.

5. 1 hr. 30 min. + 45 min.

3. 12 d. 9 hr. + 5 d. 20 hr.

6. 4 yr. 2 mo. – 1 yr. 7 mo.

Review Your Skills

1. List the units of time from longest to shortest.

sec., min., wk., d., hr.

_____ _____ _____ _____ _____
longest   shortest

6. What time will it be 45 minutes after the time shown? _____ : _____

2. Nate's workshop lasted from 9:30 A.M. to 10:45 A.M. How long was the workshop?

_____ hour(s) _____ minute(s)

9:17

7. The time is _____ minutes after _____ o'clock.

3. Circle the letter of the largest measurement of time:

a) 2 min. 10 sec.

b) 140 sec.

c) 1 min. 65 sec.

8. Write these dates using numbers.

a) June 12, 1994 ___/___/___

b) November 26, 1960 ___/___/___

9. The ninth month of the year is

_____ .

4. The baseball game is televised at 7:00 P.M. It is now 4:15 P.M. How long before the game starts?

_____ hours _____ minutes

10. Write the time 2 different ways.

a) _____ : _____

b) _____ minutes to _____

5. What time is shown on the watch?

_____ : _____

Travel Times

```
                                Bus Schedule

            ┌──────┐         ┌───────┐
            │leaves│         │arrives│
            └──────┘         └───────┘
               Lv.             Ar.                        Lv.          Ar.
  Number     Munson          Vernon        Number       Vernon        Alma

   251       9:30 A.M.      11:45 A.M.        252      12:30 P.M.    1:20 P.M.
```

▶ Use the bus schedule to answer the questions.

1. How long is the bus ride from Munson to Vernon? _____ hour(s) _____ minutes

2. How long is the bus ride from Vernon to Alma? _____ minutes

3. How much longer is the ride on bus 251 than the ride on bus 252?
_____ hour(s) _____ minutes

4. If you start at Munson and ride to Alma, how long is the wait between buses?
_____ minutes

```
                               Train Schedule

               Lv.             Ar.                        Lv.          Ar.
  Number      Dublin          Hobart        Number       Hobart       Holland

   965       1:40 P.M.       3:05 P.M.        966       4:15 P.M.    5:55 P.M.
```

▶ Use the train schedule to answer the questions.

5. How long is the train ride from Dublin to Hobart? _____ hour(s) _____ minutes

6. How long is the train ride from Hobart to Holland? _____ hour(s) _____ minutes

7. How much longer is the ride on train 966 than the ride on train 965?
_____ minutes

8. If you depart from Dublin and ride to Holland, how long is the wait between
trains? _____ hour(s) _____ minutes

Measurements with Money

1. Fund Drive

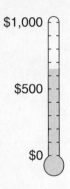

$1,000

$500

$0

a) About how much money has been collected to date?

_____

b) About how much more money is needed to reach the goal of $1,000?

_____

4. Kurt earned $9.38 per hour. How much did he earn in $8\frac{1}{2}$ hours?

_____

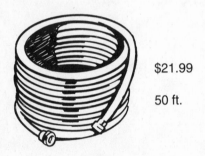

$21.99

50 ft.

2. How much will 100 feet of hose cost?

_____

5. Gasoline costs $1.56 per gallon. How much will 12 gallons cost?

_____

1 dozen = 12

$.92
per dozen

3. How much would 6 eggs cost? _____

6.

How much change will you get back from $5.00 if you buy 1 cup of hot chocolate?

_____

Hot Chocolate $.79

MEASUREMENT REVIEW

Speed Limit	
55 _____	per hour

1. Circle what goes on the line.

    a) feet   **b)** seconds   **c)** miles

2. School is out at 3:40 P.M. About how many minutes until school is out? _____

3. About how much water does the teakettle hold?

    a) 1 cup   **b)** 1 quart   **c)** 1 gallon

Bus Schedule		
	Lv.	Ar.
Number	**Sidney**	**Remus**
345	9:15 A.M.	1:05 P.M.

4. How long does it take to ride the bus from Sidney to Remus?

    _____ hours _____ minutes

5. A carpenter wants to cut 1 ft. 3 in. from an 8-ft. board. How long is the remaining board? _____

$.63 per foot

6. How much will 3 yards 2 feet of ribbon cost? _____

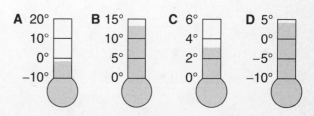

7. Denise bought 3 lb. 4 oz. of roast beef and 1 lb. 14 oz. of ham. How much meat did she buy altogether?

    _____ pounds _____ ounces

8. Estimate the temperature readings.

    a) _____   **c)** _____

    b) _____   **d)** _____

ANSWER KEY

Page 1: Scales in Everyday Life
1. A   **5.** B   **8.** H
2. C   **6.** G   **9.** E
3. I   **7.** F   **10.** J
4. D

Page 2: Reading Scales
1. a) 10   **3. a)** 5   **5. a)** 25
   b) 70   **b)** 16   **b)** 85
2. a) 2   **4. a)** 10
   b) 11   **b)** 16

Page 3: Estimate Using Scales
1. a) 6 and 7   **4. a)** 12 and 14
   b) 7   **b)** 14
   c) 7   **c)** 14
2. a) 6 and 8   **5. a)** 28 and 32
   b) 6   **b)** 28
   c) 6   **c)** 28
3. a) 30 and 35 **6. a)** 12 and 14
   b) 30   **b)** 14
   c) 30   **c)** 14

Page 4: Reading Different Scales
1. About 350   **5.** About 157
2. About 150   **6.** About 106 or 108
3. About 5   **7.** About 840
4. About 505   **8.** About 450

Page 5: Scales with Decimals
1. Start at zero.

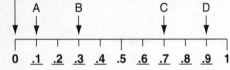

a) .1   **c)** .7
b) .3   **d)** .9

2.

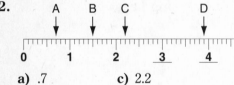

a) .7   **c)** 2.2
b) 1.5   **d)** 3.9

3.

a) .2   **c)** 1.5   **e)** 3.1
b) 2.3   **d)** .6   **f)** 1.9

Page 6: Reading Scales with Fractions
1. a) 4   **3. a)** 8
   b) $\frac{1}{4}$ **b)** $\frac{1}{8}$
   c) $11\frac{3}{4}$ **c)** $1\frac{3}{8}$
2. a) $\frac{1}{8}$ **4. a)** $\frac{1}{4}$
   b) $22\frac{7}{8}$ **b)** $5\frac{1}{4}$

Page 7: Everyday Scales
1. a) 99   **3.** About 10 miles
   b) 84   **4.** C
   c) Answers will vary.   **5.** About 75 degrees
2. About 48 mph

Page 8: Review Your Skills
1. About 63 mph   **4.** 150
2. a) 10   **5. a)** A
   b) 2   **b)** B
   c) 16   **c)** C
3. a) 1   **6. a)** .4, $\frac{4}{10} = \frac{2}{5}$
   b) $4\frac{1}{2}$ **b)** .6, $\frac{6}{10} = \frac{3}{5}$
   c) $3\frac{1}{2}$ **c)** .1, $\frac{1}{10}$

Page 9: Measuring Air Pressure
All reasonable answers are acceptable.
1. a) About 46 pounds   **3. a)** About 38 pounds
   b) About 65 pounds   **b)** About 30 pounds
   c) About 10 pounds   **c)** About 2 pounds
2. a) About 36 pounds
   b) About 55 pounds
   c) About 7 pounds

Page 10: Nonstandard Units
Answers will vary since the size of each person's thumb will be different.
1. About 7   **3.** Answers will vary.
2. About 5

Page 11: Standard Units of Length
1. d   **3.** b
2. a   **4.** b or c

Page 12: Estimate the Lengths
1. a) 2 and 3 inches   **3. a)** 1 and $1\frac{1}{2}$ inches
   b) 3 inches   **b)** 1 inch
   c) 3 inches   **c)** 1 inch
2. 1 inch   **4.** $2\frac{1}{2}$ inches

Page 13: Measuring Real-Life Objects

1. About 1 inch
2. About 1 inch
3. About $2\frac{1}{2}$ inches
4. About $\frac{1}{2}$ inch
5. About $2\frac{1}{2}$ inches
6. About $\frac{1}{2}$ inch

Page 14: Measuring with a Ruler

1. About $4\frac{1}{2}$ inches
2. About 3 inches
3. About 4 inches
4. About $1\frac{1}{2}$ inches
5. About $2\frac{1}{4}$ inches
6. About $3\frac{3}{4}$ inches
7. About $4\frac{1}{2}$ inches
8. About $\frac{1}{4}$ inch

Page 15: Draw the Measurements

1. $4\frac{1}{2}$ inches
2. $1\frac{1}{2}$ inches
3. $3\frac{1}{2}$ inches

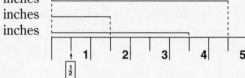

4. $3\frac{3}{4}$ inches
5. $4\frac{1}{4}$ inches
6. $2\frac{1}{2}$ inches

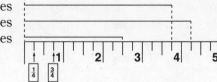

7. $3\frac{3}{4}$ inches
8. $1\frac{3}{8}$ inches
9. $4\frac{7}{8}$ inches

10. $4\frac{7}{16}$ inches
11. $2\frac{3}{4}$ inches
12. $4\frac{1}{4}$ inches

Page 16: Measuring Inches and Centimeters

1. b
2. c
3. 3
4. 76
5. 7.6
6. $1\frac{1}{2}$
7. 38
8. 3.8
9. $2\frac{3}{4}$
10. 70
11. 7
12. $\frac{3}{4}$
13. 19
14. 1.9

Page 17: Traditional and Metric Measurements

1. $\frac{3}{4}$ in.
2. $\frac{1}{2}$ in.
3. $\frac{5}{8}$ in.
4. C
5. A
6. B
7. 11 mm
8. 15 mm
9. 21 mm
10. E
11. F
12. D

Page 18: Converting Length Measurements

1. 24 inches
2. 15 feet
3. 108 inches
4. 72 inches
5. 21 feet
6. 68 inches
7. 47 inches
8. 14 feet
9. 221 inches
10. 87 inches

Page 19: More Conversions

1. 4 feet
2. 3 feet
3. 5 yards
4. 2 yards
5. 9 feet
6. 3 yards 1 foot
7. 1 yard 2 feet
8. 3 feet 4 inches
9. 1 foot 7 inches
10. 4 yards 1 foot

Page 20: Rename the Measurement

1. a) divide
   b) 4 feet
2. a) divide
   b) 2 yards
3. a) multiply
   b) 10,560 feet
4. a) multiply
   b) 48 inches
5. a) multiply
   b) 9 feet
6. a) divide
   b) 5 yards
7. 1 yard 10 inches
8. 1 foot 5 inches
9. 54 inches
10. 5 feet 7 inches
11. 120 inches
12. 220 inches

Page 21: Comparing Measurements

1. <
2. >
3. <
4. >
5. >
6. =
7. <
8. >
9. >
10. <
11. <
12. =

Page 22: Adding Lengths

1. 7 ft. 14 in. = 8 ft. 2 in.
2. 8 ft. 10 in.
3. 9 yd. 7 ft. = 11 yd. 1 ft.
4. 8 ft. 11 in.
5. 13 yd. 2 ft.
6. 4 ft. 24 in. = 6 ft.

Page 23: Subtracting Lengths

1. 1 ft. 10 in.
2. 1 yd. 1 ft.
3. 1 ft. 1 in.
4. 11 in.
5. 3 yd. 2 ft.
6. 5 yd. 1 ft.

Page 24: Review Your Skills

1. 3 feet, 70 inches, 2 yards
2. c
3. c
4. a) 24 inches
   b) 3 yards
   c) 3 feet
5. 4 units
6. 3 feet 4 inches
7. 4 yards
8. a) 15 inches
   b) 3 inches

Page 25: Using a Map Scale

The answers given are estimates. Student answers may vary.

1. About 250 miles
2. About 1,250 miles
3. About 625 miles
4. About 875 miles
5. About 1,125 miles
6. About 2,300 miles

Page 26: Weight

1. c
2. b
3. a
4. a
5. c
6. c

Page 27: Ounces, Pounds, and Tons

1. pounds (lb.)
2. ounces (oz.)
3. tons (tn.)
4. pounds (lb.)
5. ounces (oz.)
6. pounds (lb.)

Ounces	Pounds	Tons
Flashlight battery	Adult person	Cement mixer
Candy bar	Sofa	Trailer truck
Bar of soap	Bag of groceries	Airplane

Page 28: Reading Weight Scales

1. 1 pound 12 ounces
2. 3 pounds 6 ounces
3. a) 30 pounds
   b) 29 pounds 8 ounces
4. a) 98 pounds 12 ounces
   b) 99 pounds 5 ounces
   c) 100 pounds 3 ounces
   d) 101 pounds 0 ounces

Page 29: Weights with Fractions

1. a) $6\frac{3}{4}$ lb.
   b) $7\frac{1}{4}$ lb.
   c) $7\frac{5}{8}$ lb.
   d) $8\frac{1}{16}$ lb.
   e) $8\frac{1}{2}$ lb.
   f) $9\frac{3}{16}$ lb.
2. a) $3\frac{7}{8}$ lb.
   b) $4\frac{3}{8}$ lb.
   c) 5 lb.
   d) $5\frac{1}{2}$ lb.
   e) $6\frac{1}{4}$ lb.

3.

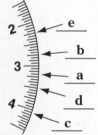

Page 30: Converting Weight Measurements

1. a) multiply
   b) 48 ounces
2. a) divide
   b) 3 pounds
3. a) divide
   b) 3 tons
4. a) multiply
   b) 160 ounces
5. a) divide
   b) 12 pounds
6. a) multiply
   b) 240 ounces
7. a) divide
   b) 6 pounds
8. a) multiply
   b) 12,000 pounds

Page 31: Rename the Measurements

1. 8 lb.
2. 192 oz.
3. 10 tn.
4. 3 lb.
5. 8 oz.
6. 1 lb. 4 oz.
7. 176 oz.
8. 72 oz.
9. 5 lb. 10 oz.
10. 67 oz.
11. 2 tn. 150 lb.
12. 4 lb. 8 oz.
13. 2 lb. 8 oz.
14. 2,450 lb.

Page 32: Estimating with Decimal and Fraction Weights

1. d
2. c
3. a
4. b

Page 33: Adding and Subtracting Weights

1. 10 lb. 5 oz.
2. 6 lb. 3 oz.
3. 10 tn. 300 lb.
4. 3 lb. 8 oz.
5. 15 lb. 1 oz.
6. 1 lb. 9 oz.

Page 34: Review Your Skills

1. About 7 lb. (Answers may vary.)
2. a) $\frac{1}{16}$ pound or 1 ounce
   b) 3 pounds 8 ounces
3. 2 lb., 35 oz., 5 tn.
4. a) 22 oz.
   b) 1 lb. 1 oz.
5. 9 pounds
6. 5 pounds
7. a) 64 oz.
   b) 9 oz.
8. a) 50 pounds
   b) 49 pounds 4 ounces

Page 35: Shipping Costs

1. $1.84
2. $5.52
3. $2.18
4. $2.76

Page 36: Cups, Pints, Quarts, and Gallons

1. a
2. c
3. b
4. b
5. a

Page 37: Fluid Ounce

1. a) $\frac{1}{2}$
   b) 4
2. a) $\frac{1}{3}$
   b) 3
3. a)
   b) $\frac{3}{4}$
4. a)
   b) 4

Page 38: Liter and Milliliter

1. a) About 120 ml
   b) About 185 ml
2. b
3. b
4. a
5. b
6. a
7. a)   b)

8. milliliters
9. milliliters
10. liters
11. milliliters
12. liters

Page 39: Converting Liquid Measurements

1. a) divide
   b) 4 pints
2. a) multiply
   b) 10 pints
3. a) multiply
   b) 20 quarts
4. a) divide
   b) 2 gallons
5. a) divide
   b) 3 quarts
6. a) multiply
   b) 14 cups

Page 40: Real-Life Conversions

1. pint
2. 16 fl. oz.
3. 3 sticks
4. 16 cups
5. 32 tablespoons
6. 8.45 servings

Page 41: Comparing Measurements

1. >
2. <
3. <
4. <
5. =
6. <
7. >
8. =
9. >
10. <
11. <
12. >

Page 42: Adding and Subtracting Liquid Measurements

1. 6 gal.
2. 3 qt. 1 pt.
3. 5 c. 2 fl. oz.
4. 3 qt.
5. 11 qt. 1 pt.
6. 3 qt.

Page 43: Review Your Skills

1. a) fl. oz., c., pt., qt., gal.
   b) ml, l
2. a) 4
   b) 8
   c) 16
3. a) gallon
   b) cup
4. c
5. quart
6. a) 6
   b) 14
   c) 2
7. b
8. a) 4 quarts
   b) 3 pints

Page 44: Measuring Ingredients in Recipes

1. Fill up cup A twice.
2. Fill up teaspoon G and H, or fill up teaspoon H three times.
3. Fill up cup C twice.
4. Fill up tablespoon E five times.
5. Fill up cup A once and cup D once.
6. Fill up tablespoon E twice.
7. Fill up teaspoon H once.
8. Fill up cup A four times.

Page 45: Real-Life Applications

1. Answers will vary.
2. 11 fluid ounces
3. 3 cups
4. $\frac{1}{2}$ gallon
5. 32 servings
6. a) $\frac{1}{4}$ cup
   b) 1 cup
   c) $1\frac{1}{2}$ cups
7. $9.15
8. 2 cups
9. 8 tablespoons

Page 46: Temperature

All reasonable answers are acceptable.
1. a) C   b) About 8°
2. a) F   b) About 38°
3. a) E   b) 68°
4. a) B   b) About 85°
5. a) D   b) 550° or Broil
6. a) A   b) 99.4°

Page 47: Fahrenheit Temperatures

1. b
2. a
3. c
4. a
5. d
6. 62°, above freezing
7. 27°, below freezing
8. 31°, below freezing
9. 33°, above freezing
10. 73°, above freezing

Page 48: Compare Fahrenheit and Celsius Temperatures

1. a) 0, 32
   b) 100, 212
2. a) 100
   b) 180
3. 68
4. 31
5. 37
6. 118
7. 82
8. b
9. b
10. a

Page 49: Temperatures Below Zero

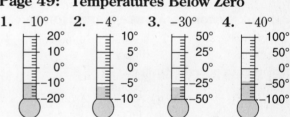

1. −10°   2. −4°   3. −30°   4. −40°

5. a) −5
   b) 18°
   c) 5°
   d) 7°
6. a) 8°
   b) −2°
   c) −4°
   d) −12°

Page 50: Rising and Falling Temperatures

1. 11°
2. 7°
3. 13°
4. 11°
5. 18°
6. 19°
7. 41°
8. 3°
9. 5°
10. −13°
11. 0°
12. 88°

Page 51: Health Thermometer

1. a) 96.4°
   b) 100.2°
   c) 97.8°
   d) 101.6°
   e) 99°
   f) 103.4°
2. a) 2.2°
   b) .6°
   c) 2.8°
   d) 3.4°
3. a) .4°
   b) 1.8°
   c) 1.2°
   d) 3.6°

Page 52: Review Your Skills

1. **Friday**   **Saturday**   **Sunday**

    60°   60°   60°
    55°   55°   55°
    50°   50°   50°
    45°   45°   45°
    40°   40°   40°

2. c
3. About −3°
4. 180°
5. 22°
6. **a)** About −28°   **d)** About −2°
   b) About 70°   **e)** About 99°
   c) About 17°   **f)** About 114°
7. −3°

Page 53: Windchill Table

1. **a)** −18°   2. 58°
   b) 40°   3. 92°
   c) −52°

Page 54: Time

1. c   3. c   5. c
2. a   4. b

Page 55: Seconds, Minutes, and Hours

1. hours   7. seconds or minutes
2. seconds   8. hours
3. minutes or hours   9. hours
4. minutes   10. seconds
5. minutes   11–20: Answers will vary.
6. hours

Page 56: Digital Clocks

1. **a)** 45 minutes after 7
   b) 15 minutes to 8
   c) 7:45
2. **a)** 52 minutes after 1
   b) 8 minutes to 2
   c) 1:52
3. **a)** 43 minutes after 9
   b) 17 minutes to 10
   c) 9:43
4. **a)** 59 minutes after 11
   b) 1 minute to 12
   c) 11:59
5. **a)** 38 minutes after 4
   b) 22 minutes to 5
   c) 4:38
6. **a)** 6 minutes after 12
   b) 12:06

Page 57: A.M. and P.M.

1. P.M.   5. P.M.   8. 10:15 A.M.
2. A.M.   6. A.M.   9. 12:30 P.M.
3. P.M.   7. 6:20 P.M.   10. 12:00 A.M.
4. P.M.

Page 58: Weekly Appointments

1. **a)** 15 minutes after 4
   b) 4:15 P.M.

2. **a)** 15 minutes to 9
   b) 8:45 A.M.

3. **a)** 30 minutes after 12
   b) 12:30 P.M.

4. **a)** 25 minutes to 10
   b) 9:35 A.M.

5. **a)** 5 minutes after 7
   b) 7:05 P.M.

Page 59: Elapsed Time

1. 7:45 A.M.   4. 8:00 P.M.   7. 6:30 P.M.
2. 7:30 P.M.   5. 9:40 A.M.   8. 11:30 A.M.
3. 4:50 A.M.   6. 2:55 P.M.

Page 60: Count the Hours and Minutes Worked

1. 8 hours 30 minutes   3. 8 hours 30 minutes
2. 3 hours 45 minutes   4. 6 hours 50 minutes

Page 61: Time Zones

1. a) 6:00 A.M.   **3. a)** eastern
b) 5:00 A.M.   **b)** mountain
c) 4:00 A.M.

2. a) 3:00 P.M.   **4. a)** 10:00 P.M.
b) 4:00 P.M.   **b)** 4:00 P.M.
c) 5:00 P.M.
d) 6:00 P.M.
e) 4:00 P.M.
f) 6:00 P.M.

Page 62: Using a Calendar

1. 31   **6.** 2
2. 5   **7. a – d:** Compare
3. Tuesday   answers with classmates.
4. 24   **8. a)** 20
5. Tuesday   **b)** March, May, August

Page 63: Reading and Writing Dates

1. a) March 7   **2. a)** 10/5/62
b) November 21   **b)** 9/6/95
c) August 3   **c)** 7/31/84
d) May 30   **d)** 12/25/99

Page 64: Converting Time

1. a) multiply   **3. a)** multiply   **5.** 256 minutes
b) 36 mo.   **b)** 56 d.   **6.** 54 mo.
2. a) divide   **4. a)** divide   **7.** 76 wk.
b) 2 d.   **b)** 4 hr.   **8.** 58 hr.

Page 65: Adding and Subtracting Time

1. 8 hr. 10 min.   **3.** 18 d. 5 hr.   **5.** 2 hr. 15 min.
2. 4 min. 17 sec. **4.** 5 hr. 45 min. **6.** 2 yr. 7 mo.

Page 66: Review Your Skills

1. Wk. (week),   **6.** 3:20
d. (day), hr. (hour),   **7.** 17 minutes
min. (minute),   after 9 o'clock
sec. (second)   **8. a)** 6/12/94
2. 1 hour   **b)** 11/26/60
15 minutes   **9.** September
3. b   **10. a)** 4:45
4. 2 hr. 45 min.   **b)** 15 minutes
5. 4:53   to 5 o'clock

Page 67: Travel Times

1. 2 hours 15 minutes   **5.** 1 hour 25 minutes
2. 50 minutes   **6.** 1 hour 40 minutes
3. 1 hour 25 minutes   **7.** 15 minutes
4. 45 minutes   **8.** 1 hour 10 minutes

Page 68: Measurements with Money

1. a) $650   **3.** $.46   **5.** $18.72
b) $350   **4.** $79.73   **6.** $4.21
2. $43.98

Page 69: Measurement Review

1. c   **7.** 5 pounds 2 ounces
2. About 35 minutes   **8. a)** About −2°
3. b   **b)** About 13°
4. 3 hours 50 minutes   **c)** About 3°
5. 6 feet 9 inches   **d)** About 4°
6. $6.93